SOURCES AND METHODS IN GEOGRAPHY

Series editors

M.A. Morgan PhD
Department of Geography, University of Bristol

D.J. Briggs PhD
Department of Geography, University of Sheffield

SOURCES AND METHODS IN GEOGRAPHY

Historical Sources in Geography

Michael Morgan PhD
Senior Lecturer in Geography,
University of Bristol

BUTTERWORTHS
London · Boston
Sydney · Wellington · Durban · Toronto

The Butterworth Group

UK   **Butterworth & Co (Publishers) Ltd**
      London: 88 Kingsway, WC2B 6AB

Australia   **Butterworths Pty Ltd**
             Sydney: 586 Pacific Highway, Chatswood, NSW
             2067
             Also at Melbourne, Brisbane, Adelaide and Perth

Canada   **Butterworth & Co (Canada) Ltd**
          Toronto: 2265 Midland Avenue, Scarborough
          Ontario M1P 4S1

New Zealand   **Butterworths of New Zealand Ltd**
               Wellington: T & W Young Building,
               77–85 Customhouse Quay, 1, CPO Box 472

South Africa   **Butterworth & Co (South Africa) (Pty) Ltd**
                Durban: 152–154 Gale Street

USA   **Butterworth (Publishers) Inc**
       Boston: 19 Cummings Park, Woburn, Mass. 01801

First published 1979

ISBN 0 408 10609 3

Typeset & produced by Scribe Design, Chatham, Kent
Printed in England by Cox & Wyman Ltd, London, Fakenham
and Reading

British Library Cataloguing in Publication Data

Morgan, Michael Alan
   Historical sources in geography.
   — (Sources and methods in geography).
   1. Geography, Historical — Methodology
   I. Title   II. Series
   910   G141   78-40365

   ISBN 0—408—10609—3

FOREWORD

During recent years, geography has been undergoing considerable change. There have been many facets to this change, but one underlying theme is the adoption of a more rigorous approach to geographical enquiry, wherever this is appropriate. It has been reflected in numerous ways: in the greater emphasis which is placed upon quantitative and statistical methods of data collection and handling; in the attention given to the study of process as opposed to the description of form in human as well as in physical geography; and in the use of an inductive rather than deductive philosophy of learning.

What this means in practical terms is that the student and teacher of geography need to be acquainted with a wide range of methods. The student, both at school and in higher education, is increasingly becoming involved in projects or classwork which include some form of individual and original research. To be equipped for this type of study he needs to be aware of the sources from which he can obtain data, the techniques he can use to collect this information and the approaches he can take to analyse it. The teacher similarly requires a pool of empirical material on which he can draw as a source of class exercises. Both must be able to tackle geographical problems in a logical and scientific fashion, to construct appropriate explanatory hypotheses, and test these hypotheses in an objective and rational manner.

The aims of this series of books are therefore to introduce a range of sources which provide information for project and classwork, and to outline some of the methods by which this material can be analysed. The main concern is with relatively simple approaches rather than more sophisticated methods.

The reader will be expected to have a basic grounding in geography, and in some of the books a working knowledge of mathematical methods is useful. The level of detail and exposition, however, is intended to make it possible for the student, with little further reading, to gain a basic understanding of the selected themes. Consequently the series will be of particular interest and use to students and teachers involved in courses in which practical and project work figure as major components. At the same time students in higher education will find the books an invaluable guide to geographical methods.

M.A. Morgan   D.J. Briggs

ACKNOWLEDGEMENTS

The following have very kindly given me permission to reproduce the items that have been cited in the text.

Mr. M.J. Blakemore.

Prof. H.C. Darby and Mr. R.W. Finn.

Dr. J.H. Farrington.

Dr. J.D. Hamshere.

Mr. H.A. Hanley, County Archivist of Buckinghamshire.

Dr. R. Holmes.

Dr. J. Oliver.

Prof. R.F.W. Peel.

Miss P. Sammons.

Dr. G. Shaw.

Dr. T. Wild.

Dr. E.A. Wrigley.

PREFACE

The aim of this book naturally reflects the broad aims of the series as a whole, namely to introduce a range of source material, some of it relatively little used in geography; to provide information upon which individual or group project work can be based and to outline the methods by which the material can be analysed. If in this particular book the sources appear to have been stressed rather more heavily than the methods it is because, although there are many texts available to explain simple methods of statistical analysis, there is a dearth of material on historical sources for geographers. Happily there is no lack of sources themselves; indeed the contrary is the case and because of the richness of the material available some difficult decisions had to be made about which to leave out.

Since some of the sources mentioned are almost certain to be unfamiliar, where it seems appropriate, some background has been lightly sketched in, to give a context to particular sources, but this volume is not intended as a textbook on historical geography. Rather it is meant to introduce the student to certain historical sources that are relatively easily available; to show the sort of work that can be done using these sources, and in a general sense to increase interest in aspects of the geography of the past. This should open up new areas from which project themes can be drawn, so freeing students from the natural, but in human geography often limiting, emphasis on a small number of stereotyped subjects that are the staple fare of many individual studies, and sparing them the feeling that the path they tread has been beaten smooth by the feet of the many who have followed it before.

There are two broad directions in which the historical dimension is likely to be of use, and practical considerations usually mean that we choose one or another but rarely both.

One direction is towards increasing depth of treatment over time of a small area, let us say exploring over a longer and longer time span the ways in which the people and the land of a parish have interacted. Rural parishes with their long tradition of settlement and slow evolution rather than dramatic change appeal strongly to the imagination; this is reinforced by the records of the community which are often still available locally. Once interest is aroused there is great satisfaction in being able to fill in details and so build up a more accurate picture of the parish over a longer time span. This is often done by following one theme over time, for example farming or population, though the parallel strands may well be interwoven from time to time to bring out the distinctive flavour of the different periods.

The other direction is more like the 'snapshot' approach where the interest centres on a particular area at a particular time in the past and on the pattern made by the interaction of different human and/or physical elements in all their complexity at that time. So, for example, one might be interested in the Domesday geography of a group of parishes and in the factors that influenced land use patterns at this time. Or the starting point might be county-wide variations in taxable wealth in say the 1334 Lay Subsidy.

Each direction holds promise and pitfalls and progress in both is made a great deal easier if one is familiar with the great range of sources available and aware of their reliability and accuracy.

M.A.M.

CONTENTS

ILLUSTRATIONS

TABLES

CHAPTER 1 INTRODUCTION

The watershed between history and prehistory is traditionally the written word. Prehistorians have only the material remains from which to reconstruct images of early societies. With the written word, whether lists of the victories of powerful kings, palace accounts, taxation returns or poems of a more personal nature, the values, structures and economic bases of societies immediately become more accessible to the trained observer and interpreter.

In this book we shall confine ourselves to the United Kingdom and we shall start our survey of sources with Domesday. This is not of course to deny the interest of earlier periods. The centuries of Roman rule left an enduring mark on much of the country. The Dark Ages following the collapse of Roman authority in Britain in the fifth century were a turbulent period yet the rich achievements of the Anglo-Saxons included a literary tradition of high quality. However, because of the amount of accessible written material of major geographical interest included in the Domesday book, this seems by far the most useful place to begin.

The Middle Ages warrant a separate section because of the strength and importance of the rural economy for much of the time and because of the diverse nature and the great quantity of the surviving written material from that period.

It is then convenient to move from a chronological to a systematic framework and Chapter 4 deals with agriculture. It begins with the manorial farm economy and the subsequent transition from open fields to enclosure, continues with the contentious matter of tithes and tithe commutation and ends with the question of the provenance and reliability of early agricultural statistics. Population is the subject of Chapter 5. The main part of this chapter deals with the evidence for the variations in the numbers of people at parish level from the period of the parish registers in the mid-sixteenth century until the modern census, and explains the opportunities and problems presented by the varied sources on social matters, such as the accounts of the Churchwardens and the Overseers of the Poor. Chapter 6 is mainly concerned with the use of county and city directories but also mentions rate books and local newspapers. Chapter 7 deals in a cautionary way with roads, canals and railways.

Perhaps the most obvious omissions are sections dealing specifically with industry and with towns. This is intentional but does not mean that neither subject is interesting or rewarding, rather the reverse in fact. The sheer volume of material on both subjects reflects a growing interest and it is difficult in a book of this size to deal systematically with so much material. Since the type and scale of both towns and industry vary so widely from place to place and from time to time, sample studies are not as useful as they may be in other sections. On the other hand a great deal of the material in Chapter 6 is very much orientated towards urban and therefore to some extent towards industrial subjects so the themes are not wholly neglected.

CHAPTER 2 THE DOMESDAY BOOK

2.1 INTRODUCTION

The two principal volumes of the Domesday Book are one of the most important sources for the historical geographer. The Domesday Book was compiled in 1086, just 20 years after the Conquest, by the clerks of the King's Curia at Winchester and it impresses us today, as it impressed those making it, by its remarkable comprehensiveness. There is no other historical document of the period that covers so much of the country in such great detail. Only Northumberland, Durham, Cumberland and northern Westmorland were omitted. To make the survey, the country was divided up into seven or possibly nine circuits, each made up of a number of counties. Sworn jurors from each hundred, a subdivision of a county, appeared before the commissioners for each circuit. The terms of reference of these commissioners were to establish:

'.....what is the name of the estate, who held it in the time of King Edward, who holds it now, how many hides there are, how many ploughs in demesne and how many are held by the tenants, how many villeins, how many cottars, how many slaves, how many freemen, how many sokemen, how much wood, how much meadow, how much pasture, how many mills, how many fisheries, how much has been added or taken away, how much the whole was worth then and how much now......All this three times, namely in the time of King Edward, and when King William gave it and as it is today, and if it is possible that more can be obtained how it is to be obtained.' (Galbraith, 1948).

The details were apparently collated in regional centres and a summary of the returns for each circuit was sent to Winchester, where the Royal Treasury clerks transcribed them into what we now know as Volume I of the Domesday Book, the so-called Exchequer Domesday. Volume 2, the Little Domesday relates to the East Anglian circuit which was never, for some reason, recast in the final form to make a part of Volume I. The other surviving Domesday document is the Exeter Domesday, which is now known to be the first unordered draft of the returns for the southwestern circuit, covering Devon, Cornwall, Somerset, Wiltshire and Dorset. The advantage of the Exeter version is that it gives more details, particularly of livestock, than the Exchequer Domesday.

Virtually all the Domesday Book entries have been translated and are to be found in the Victoria County Histories. The original survey was made in terms of manors and hundreds, but the clerks at Winchester rearranged the entries under the heading of the individual land holders in each county, beginning in each case with the royal manors. If one is looking for a particular parish or manor it is essential to use the index, as it is quite possible that if a community is made up of two different but contiguous manors the entries for each will appear in quite different parts of the record. In the southwestern counties a comparison between the Exeter and the Exchequer version often reveals some innaccuracies caused by careless copying. This means that one has to be prepared for some errors in the Exchequer version though one is usually unable to check from other sources. It has also transpired that some of the identifications of Domesday places made by the compilers of the Victoria County Histories are wrong, so caution is needed here as well.

If the place in which one is interested is not mentioned in Domesday, this may be because it did not then exist; or it may be that it is in Domesday but under another name. Sometimes it is known that a particular place was named in Anglo-Saxon charters, or that the name is so ancient that the place had to be in existence at the time of Domesday, even if for some reason it is not listed.

2.2 FORMAT

Some idea of the nature of Domesday as a source may be gained from the extracts for the village of Newton St Loe which lies a few miles to the west of the city of Bath in Somerset. This is the entry taken from the Victoria County History for Somerset, based on the fuller Exeter version:

> The Bishop (of Coutances) himself holds NIWETONE. Alvric held it T.R.E. and paid geld for 3 hides. There is land for 4 ploughs. In demesne are 2½ hides and 2 ploughs and 4 slaves and there are 4 villeins and 3 bordars with 2 ploughs and half a hide. There are 2 riding horses and 12 beasts and 40 swine and 39 sheep. There is a mill paying 7/6d. and 9 acres of meadow and 40 acres of underwood. It was worth 60/-, now 100/-.
>
> To this manor have been added 7 hides which 9 thegns held in parage T.R.E. There is land for 8 ploughs. In demesne are 3½ hides and 1½ virgates. There are 14 villeins and 8 bordars and 7 slaves with 6 ploughs and 13 acres of meadow. It was worth 100/-, now £10.

This entry can be interpreted as follows:
'The Bishop of Coutances' was from Normandy, and was greatly favoured by King

William who gave him many manors in this part of the country. The little town of St Loe is also in Normandy and since the Conquest the name has attached to the village of Newton.

'Alvric' was the name of a Saxon thegn or a minor nobleman, whose lands probably were taken by the victorious Normans.

'T.R.E.' is short for the Latin, Tempore Regis Edwardi (in the time of King Edward, i.e. in 1066 immediately before the conquest).

'Geld' is a form of tax on land dating from before the Conquest, but continued for some time afterwards.

'Hide' is a measurement of area. In this part of the country it is thought to have been roughly equal to 120 acres.

'Land for 4 ploughs' means that the arable land would need four teams of plough oxen or other draught animals to cultivate it.

'Demesne land' is land in the direct ownership and operation of the lord of the manor. Generally it made up between ½ and ⅓ of the total and like the peasant holdings was scattered about in strips over the common fields that usually surrounded the village.

'Villeins' were on the highest rung of the social ladder amongst the peasantry and were the most numerous at the time of Domesday. A villein normally held a virgate (a quarter of a hide and therefore about 30 acres of land), also in strips in the common fields. He had to provide two oxen for the common plough and he was obliged to work from two to three days a week throughout the year on his lord's demesne. For this he received payment in kind, hens at Christmas, eggs at Easter, grain at Martinmas, and also payment in money. At the time of Domesday a villein was in many respects not very different from a freeman: his landlord could not eject him from his holding as long as he performed the services due for it. Below the villeins were the bordars and cottars, holding less land and having to render less service to the lord. Having no cattle, for example, they were not required to give ploughing service. They also seem to have had less security of tenure. Then came the coliberts who along with the slaves formed the lowest group of all. Coliberts were, in many respects, little better off than slaves. They had no land and were wholly dependent on working on the demesne. In time the slave class died out and was replaced by cottars who were found to be both cheaper and more efficient.

'A mill paying 7/6d.' i.e. the tax levied on the mill. It could have been a wind mill or a hand mill or as in the case of Newton a water mill.

'Underwood' was of significance both as a source of firewood and as pannage for swine, providing beech mast, nuts and berries.

'Was worth 60/-, now 100/-.' These figures represent the taxes raised in 1066 and 1086 respectively. Here obviously there was an increase since the Conquest. Many manors in other parts of the country, especially in the North, show a marked decrease in this period, reflecting the depredations wrought by the Norman armies as they overcame resistance.

'To this manor have been added' This gives a strong suggestion of two separate groups of settlements, borne out by other late medieval evidence, having been brought together after the Conquest.

'In parage', means held jointly, with equal responsibility, a form that obviously did not survive the Conquest in this area.

2.3 INTERPRETATION

Since Victorian times a great deal of scholarly effort has been devoted to interpreting Domesday, yet despite this a number of important questions remain unanswered or cloaked in controversy. Hoskins (1972) warns: 'It is quite true that simple answers to difficult questions about the meaning of Domesday Book are likely to be misleading. There are so many questions arising from this record that seem to be simple to the local historian who is asking them but can only be answered with a number of qualifications in the light of special local circumstances'.

The most significant geographical interpretation to date is the five volume *Domesday Geography of England* edited by Darby (1952–1967). Before looking at any individual manor or small area it is essential to consult the appropriate volume of Darby in order to get a broad perspective. An example of the sort of detail of spatial patterning that is revealed by Darby's methods is shown in *Figure 2.1.*

Even if one is mainly interested in a single manor it makes sense to examine the entries for a number of neighbouring manors so that one can see how far, if at all, this manor is typical of those in the district. For comparative purposes the old hundred divisions of the counties make a useful unit.

2.3.1 Problems

Table 2.1 summarises the Domesday entries for all the settlements in the Hundred of Wellow, which includes Newton. It illustrates some of the problems encountered when the material is looked at in really close detail.

One problem arises immediately when one tries to identify the Domesday manors. The village of Wellow for example gives its name to the Hundred, so it must have been important,

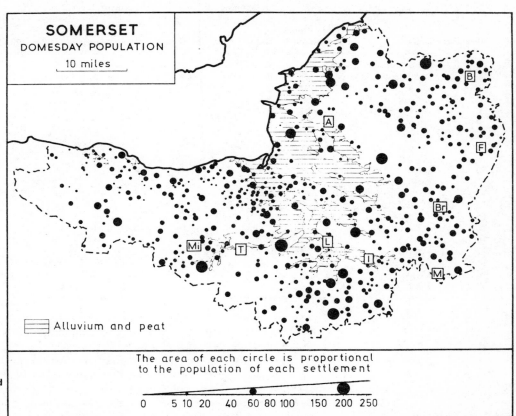

Figure 2.1 Somerset: Domesday population in 1086 (by settlements). Domesday boroughs are indicated by initials: A, Axbridge; B, Bath; Br, Bruton; F, Frome; I, Ilchester; L, Langport; M, Milborne Port; Mi, Milverton; T, Taunton (From Darby and Finn, 1967, *Domesday Geography of South West England*, courtesy of Cambridge University Press)

SOMERSET
DOMESDAY POPULATION
10 miles

Alluvium and peat

The area of each circle is proportional to the population of each settlement

0   5 10 20   40 60 80 100   150 200 250

TABLE 2.1 DOMESDAY CHARACTERISTICS OF SETTLEMENTS IN THE HUNDRED OF WELLOW, SOMERSET

Reference	Name	Geld paid	Plough-lands	In demesne			Peasantry				Beasts	Swine	Sheep	Meadow
				Hides	Ploughs	Slaves	Villeins	Bordars	Hides	Ploughs				
1	Camerton	10	10	7	2	8	6	6	2	2	—	13	154	80
2	Carlingcott	3½	3	3¼	2	1	1	3	¼	—	3	12	162	10
3	Corston	10	9	5	2	4	5	8	5	3	—	6	62	6
4	Dunkerton	3	8	1½	4	8	10	6	1½	4	11	36	212	6
5	Englishcombe	10	10	6¾	3	6	3	19	4	6	9	24	137	—
6	Evestie	1	—	—	1	3	—	—	—	—	—	—	—	4
7	Farleigh	½	—	1⅛	—	—	1	5	⅛	1	—	3	—	3
8	Foxcote	5	4	3¾	2	3	3	8	1	2	20	29	177	19
9	Hinton	10	10	5	3	9	12	15	5	6	40	40	250	12
10	Newton St Loe	3	4	2½	2	4	4	3	½	2	12	40	93	9
11	Newton St Loe	7	8	3⅝	—	7	14	8	3⅜	6	—	—	—	23
12	Norton St Philip	10	10	5¼	3	3	3	16	5	3	20	20	240	20
13	Stony Littleton	2	2	—	2	7	—	—	—	—	5	15	200	2
14	Tellisford	2	3	1½	2	0	—	9	½	—	7	13	95	7
15	Tellisford	3	4	1¾	1	3	3	8	—	2	—	12	65	11½
16	Twerton	7½	10	3½	3	6	7	13	3	6	11	17	200	15
17	Twerton	2½	2½	2½	2½	6	—	—	—	—	6	18	200	7
18	Whittoxmeade	1	2	⅝	2	1	—	6	⅜	—	1	12	120	3
19	Wick	¼	1	—	—	1	1	—	—	—	—	—	—	—
20	Woodborough	1	2	⅛	1	7	—	—	—	—	—	—	150	—

yet it does not appear in Domesday. However, the modern parish of Wellow includes the settlements of Eckweek, Stony Littleton, Woodborough and White Ox Mead, and all of these are recorded as separate holdings in the Domesday record.

The hideage on which geld is paid is given first for the whole manor, and then a separate figure is usually given for the hideage in demesne and the hideage operated by the peasants for themselves. The subtotals ought to tally with the total figure but they frequently do not.

Similarly one would expect that the total under ploughlands would equal the sum of the ploughs recorded under demesne and under peasant ownership, but for example in the whole of Somerset the figures balance in only a third of the total entries. Half of the entries record a deficiency of ploughteams, and only a small number show a surplus. One might be tempted to suppose that where there was a deficiency of ploughteams over ploughlands some crisis had occurred to leave once arable land uncultivated. If this were indeed the case it ought also to

TABLE 2.1 continued

Pasture	Value (£)		People recorded	Ratio Ploughlands: Ploughs
	1066	1086		
20	6	7	20	10:4
—	3	2½	5	
—	7	8	16	9:5
80	5	6	24	8:8
12	10	10	28	10:9
—	—	1	3	0:1
—	1	½	6	0:1
6	4	4	14	4:4
—	10	12	36	10:9
—	3	5	11	4:4
—	5	10	29	8:6
20	6	7	22	10:6
6	1½	2	7	2:2
10	½	1	9	3:2
30	3	2	14	4:3
—	10	10	26	10:9
—	3	3	6	2½:2½
—	3	3	7	2:2
—	½	½	2	1:0
—	1½	2	7	2:1

2.3.2 Population

follow that the value of the manors where it had happened fell between 1066 and 1086. Yet in Somerset at least as a whole there was no consistency in this relationship. The value of deficient holdings sometimes went down, but sometimes it remained unchanged, and in a few cases it even rose. In Wellow Hundred in most cases there was a deficiency of ploughs in relation to ploughlands though commonly it was not a very large one and there seems no obvious relationship between this and the changes in value.

Changes in value are hard to account for and in most cases so many factors are involved, the explanation will never be known. It does seem probable that in this part of Somerset the changes are most likely to have been a result of natural causes, economic circumstances or even purely administrative changes, rather than, as in the North, of military action.

Generally, the greater the number of ploughteams on a manor the greater was its value as one would expect considering the close relationship between population and ploughteams. This is clearly brought out in *Figure 2.2.* Other factors however, such as the amount of meadow and pasture, of woodland and livestock would also have an effect so one should not jump to conclusions about the relationship. There are also cases which are difficult to account for where ploughs and ploughteams are recorded but not peasantry.

While there is often a correlation between high population and high ploughteam density, this is by no means universal. East and Central Hertfordshire with a lower population density than Central Leicestershire appear to have had a higher ploughteam density. This should warn us against assuming that a high population density necessarily indicates a high level of prosperity. In such areas there may have been many peasant families without ploughteams, or with only a very small share in such teams.

The entry for each manor mentions the number of peasants, all of whom in their different ways were important in the feudal hierarchy. The totals in each manor in Wellow Hundred are recorded in *Table 2.1.* With no less than 40 in the two manors Newton St Loe has the greatest number in relation to the ploughlands and is one of the most populous manors. Obviously this figure of 40 is not a complete count of the population. Wives, children, manorial officials and members of the Lord's household and family are not recorded. Bearing in mind that we are considering a period almost a thousand years ago it might be thought idle to speculate about how many individuals the manor actually supported. But in fact a very careful sifting of what scraps of evidence have come to light from different parts of England and Western Europe suggests that multiplying the numbers of individuals recorded by 3.5 will give a reasonable approximation of the total population.

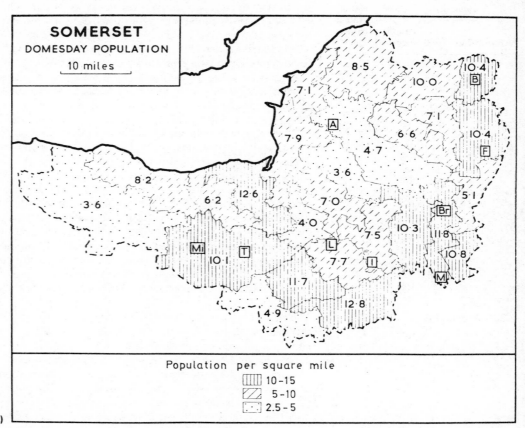

**Figure 2.2 Somerset: Domesday
population in 1086 (by densities). See
Figure 2.1 for key to initials (From
Darby and Finn, 1967,** *Domesday
Geography of South West England*,
courtesy of Cambridge University Press)

SOMERSET
DOMESDAY POPULATION
10 miles

Population per square mile

‖‖‖ 10–15
⧄ 5–10
⋯ 2·5–5

8·5   10·0   10·4 B
7·1   10·0   7·1   10·4 F
A   7·9   6·6
7·9   4·7
3·6   5·1
8·2   6·2   12·6   7·0   Br
3·6   4·0   7·5   10·3   11·8
Mi T   L   7·7   II   10·8 M
10·1   11·7   12·8
4·9

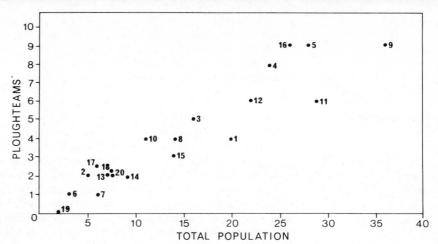

Figure 2.3 Relationship between recorded Domesday population and the number of ploughteams, Hundred of Wellow, Somerset. Numbers relate to the reference numbers used in Table 2.1

If one feels some reservations about applying a multiplier of the same value in all cases there is no need to use this method. In comparing the situation over an area there is no reason why one should not use the raw figures. Assuming that they were compiled on a common basis the position of each manor relative to all the others will be preserved.

Where it is possible to work on a number of Domesday manors it is well worth considering ways of exploring relationships between the variables recorded in the survey. Examples of some relationships are shown in *Figure 2.3* which attempts in a relatively unsophisticated way to indicate the general nature of the relationship rather than to provide a statistic, and it may well be worth attempting this for other hundreds. One should not be surprised if exercises like this raise more questions than they answer. It is also important to make sure that the number of manors in any hundred is sufficient for the method being used.

In many parts of the country the Domesday Book shows that the use made of the land, the pattern of its peopling, bears some relationship to influences such as the quality of the soil, the topography and the extent of forest and woodland, though we must avoid the temptation to be too deterministic. Darby's work has already shown the broad effects of physical conditions and most of the maps hint at significant relationships between land use and, for example, soil. At the present time it is worth trying the same method for Domesday, taking care to select a group of manors that runs across strongly contrasted country. Naturally care is needed not to read too much into an apparent relationship the moment it appears to suggest itself. Should it seem that some significant relationship is emerging then further transects will be needed to confirm or refute it.

2.3.3 Contemporary computer analysis

It will soon become apparent after only a brief acquaintance with the Domesday material that the sheer weight of data is overwhelming. Until recently the painstaking plotting of the data for an area of any extent to show significant spatial patterns and the linking of these patterns with certain elements of the physical environment seemed the best that could be hoped for. Certainly the Domesday geography of England that emerged under Darby's inspiration has helped greatly to establish patterns and to suggest causes. But the labour that is involved is immense. However the computer can handle quantities of data that are well beyond the scope of manual methods, and a preliminary indication of the possible results of analysing Domesday data has recently been given by Hamshere and Blakemore (1976). They took seven counties selected from the Exchequer Domesday, giving a broad transect from Middlesex in the east to Worcestershire in the west. By so doing they included more than one of the Domesday circuits, but avoided the problems presented in the west by the more detailed Exeter version and in the east by the different format of Volume 2.

There are difficult problems even when one uses the superior capacity of the computer. One difficulty arises because of multicollinearity of the data. That is, it is hard to decide from the data which variables are dependent and which independent. An example of this would be the three variables: (a) 'area cultivated' (effectively ploughlands); (b) 'plough-teams'; and (c) 'population'. Each is likely to be dependent to some extent on the others, and where this is so, classical regression analysis can only be used with sophisticated procedures that improve the acceptability of the result to some degree. Using such methods that are beyond the scope of simple hand methods Hamshere and Blakemore derived *Table 2.2*

One can see immediately that in all seven counties there is a consistent and significant relationship between total rural population and the number of ploughteams, as indeed one might expect in an area dominated by the great open fields of the Midland farming system. Indeed we found the same relationship in Wellow (*Figure 2.4*) using less sophisticated methods.

TABLE 2.2 CORRELATION COEFFICIENTS: PLOUGHTEAMS AND POPULATION (after Hamshere and Blakemore, 1976)

County	Total population (minus burgesses)	Villeins	Bordars	Cottars	Slaves	Number of Domesday entries
Middlesex	0.95	0.96	0.622	0.54	0.52	99
Hertfordshire	0.92	0.93	0.80	0.68	0.82	331
Buckinghamshire	0.95	0.95	0.80	—	0.79	391
Huntingdonshire	0.94	0.95	0.73	—	—	150
Oxfordshire	0.95	0.93	0.84	—	0.74	397
Warwickshire	0.95	0.94	0.73	—	0.69	365
Worcestershire	0.94	0.88	0.84	0.11	0.67	233
Transect total	0.93	0.90	0.78	na	0.69	1966

TABLE 2.3 CORRELATION COEFFICIENTS: DEMESNE PLOUGHTEAMS AND POPULATION (after Hamshere and Blakemore, 1976)

County	Total population (minus burgesses)	Villeins	Bordars	Cottars	Slaves
Middlesex	0.86	0.70	0.81	0.27	0.73
Hertfordshire	0.85	0.86	0.73	0.66	0.84
Buckinghamshire	0.88	0.85	0.76	—	0.82
Huntingdonshire	0.81	0.81	0.62	—	—
Oxfordshire	0.88	0.83	0.79	—	0.81
Warwickshire	0.85	0.79	0.69	—	0.77
Worcestershire	0.80	0.73	0.61	0.12	0.82
Transect total	0.84	0.77	0.70	na	0.77

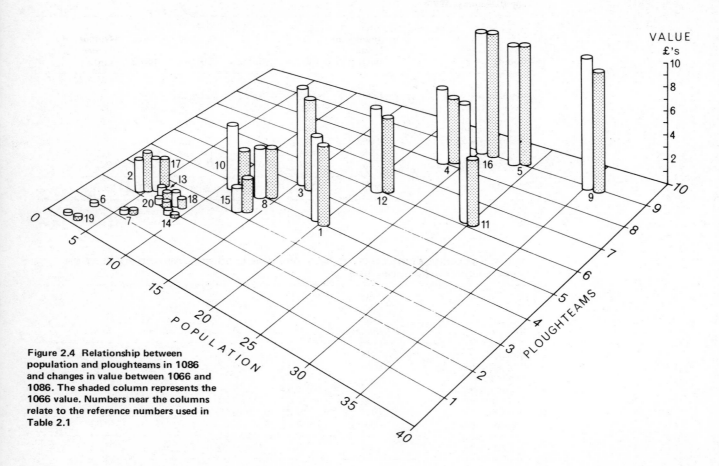

Figure 2.4 Relationship between population and ploughteams in 1086 and changes in value between 1066 and 1086. The shaded column represents the 1066 value. Numbers near the columns relate to the reference numbers used in Table 2.1

TABLE 2.4 CORRELATION COEFFICIENTS: PEASANT PLOUGHTEAMS AND POPULATION
(after Hamshere and Blakemore, 1976)

County	Total population (minus burgesses and slaves	Villeins	Bordars	Cottars	Slaves
Middlesex	0.91	0.96	0.53	0.52	0.44
Hertfordshire	0.90	0.92	0.80	—	0.78
Buckinghamshire	0.94	0.94	0.77	—	0.74
Huntingdonshire	0.94	0.95	0.73	—	—
Oxfordshire	0.93	0.92	0.81	—	0.67
Warwickshire	0.94	0.94	0.70	—	0.64
Worcestershire	0.92	0.86	0.83	0.10	0.60
Transect total	0.90	0.89	0.76	na	0.63

The association is almost as strong when one takes villeins only as a group with the possible exception of Worcestershire (0.88). However when demesne ploughteams are distinguished separately Middlesex stands out as different, with the bordars more significantly correlated (0.81) with demesne teams than villeins (0.70) (*Table 2.3*)

In *Table 2.4* the coefficients between peasant ploughteams and the various groups make it clear that for Middlesex more than any other county the villeins had the highest correlation (0.96) and the slaves the lowest (0.44). A possible explanation of this could be that the villeins of Middlesex included a number of the more independent peasantry who had become depressed into the peasant class after the Conquest, though at this stage this can be no more than an interesting theory that needs further investigation. Without a computer facility it is unlikely that this sort of inference could have been made. Another advantage is that the result of a computation can be almost instantly mapped using computer-linked plotting equipment. This allows one to speculate about possible relationships from a set of maps and to test any ideas very quickly indeed. This process, called interactive mapping, is likely to develop considerably. As far as Domesday is concerned the geographical interpretation has tended so far to emphasise the role of the physical environment in accounting for differences in land use and economy. It may be that when Domesday data are analysed in other ways other factors, besides physical environment, may prove to be equally or more important in accounting for spatial differences. It may be, for example, that if all the land owned directly by the church is compared to that in lay hands, or to that in Royal demesne, differences will appear as great as or greater than those attributable to the physical environment. If the attempt to analyse the entire Domesday book by the 900th anniversary of its compilation (1986) is successful we shall probably by then know as much as ever we shall about the economy and society of that far-off period.

REFERENCES

Darby, H.C. (Ed.) (1952–1967). *The Domesday Geography of England*, 5 vols. Cambridge University Press

Galbraith, V.H. (1948). *Studies in the Public Records*. London; Nelson

Galbraith, V.H. (1961). *Making of the Domesday Book*. Oxford University Press

Hamshere, J.D. and Blakemore, M.J. (1976). 'Computerising Domesday Book.' *Area* **8** pp. 289–294

Hoskins, W.G. (1972). *Local History in England*, 2nd Edn. Harlow, Essex; Longman

CHAPTER 3 THE MIDDLE AGES: DOMESDAY UNTIL THE DISSOLUTION OF THE MONASTERIES

3.1 INTRODUCTION

The Domesday Book presents a unique picture of most of England at the time of the Conquest. Throughout the Middle Ages there is no other document or series of documents of equal breadth and detail. Yet the Middle Ages was a period of great and far-reaching changes, including the breakdown of the feudal system, the growth of towns and with them of commerce and industry, and important changes in farming which however remained the mainstay of the economy.

Countless documents have survived from this period, some translated, others awaiting expert attention, and with the passage of time scholarly judgements are supplanted, sometimes reversed by new evidence or by new interpretations of familiar material. The language and style of the contemporary documents place them beyond the understanding of all but trained historians unless they have, by good fortune, been translated. Therefore, the historical geographer must either equip himself to handle original documents or at least accept the disadvantage of not being able to do so.

In practice, as far as the aims of this book are concerned, the medieval material, uneven though it is, is important for the glimpses it gives the historical geographer of what happened after Domesday and before the period when documentary and field evidence is more explicit. If one is interested in a small area, a parish or a group of parishes, it pays to comb through as many sources as possible, collecting all the information that comes to hand. Not all will be of use, but until it has all been gathered one has no way of knowing which pieces will fit into each other.

The sources we shall look at in this section are all available, at least in the major libraries, in translation, and so should prove not too difficult to handle. Matters relating most particularly to agriculture in the medieval period are discussed in Chapter 4. It is unlikely that all the sources mentioned will be of equal use or relevance in a particular enquiry. The list is intended more as a guide than as a route map.

3.2 LAY SUBSIDIES, POLL TAXES AND INQUISITIONS

3.2.1 The 1334 Lay Subsidy

Medieval Lay Subsidies were taxes levied on the personal wealth of the laity when the

English Crown needed more money, especially to offset the cost of wars. At first assessments were made on each individual's wealth but by 1334 a new system was introduced whereby each vill and borough was allotted a quota of tax and it had to decide how to apportion the responsibility for providing it. The 1334 Lay Subsidy is especially important because it involved nearly 14 000 places all over England in the decades prior to the Black Death. The 1334 Subsidy has been edited, transcribed and the settlements located and identified by Dr R. Glasscock (1975). The lists taken as a whole give an almost complete picture of variations in wealth over the country. A zone of wealth can be picked out stretching southwest to northeast from Somerset and Wiltshire to Norfolk, Lincolnshire and the East Riding of Yorkshire. The eastern part of Kent, the coastal plains of Hampshire and Sussex and the Thames Valley were up to three times as prosperous as the West Midlands and the North at that time.

The 1334 Lay Subsidy is unrivalled as a source of national data but there were many other such subsidies at different periods, either covering smaller areas or only partly surviving, and there were also Papal Inquisitions designed to assess the contributions that parishioners could be expected to make for various purposes. Local record societies have in many cases published translations of some of these surveys where they are of interest and these are of great value to the historical geographer. As with the Domesday Survey the returns for a single settlement often gain in significance for the geographer if they can be compared with other places in the district.

3.2.2 The 1327 Lay Subsidy

The Exchequer Lay Subsidy return of 1327 for the manor of Newton St Loe is shown in *Table 3.1*. The names alone are of interest, indicating as they do a mixture of French and English origins. Many of the names recorded here persist in the parish well into the period covered by the registers of baptisms, marriages and burials from the sixteenth century, indicating a certain degree of stability in the population. *Table 3.2* gives the summary totals for the Hundred of Wellow and shows a fair amount of variation of assessment and so presumably of wealth within quite a small area. It is interesting to compare the relative standing of the parishes in 1327 and 1086.

3.2.3 The Lay Poll Tax returns of 1377, 1379 and 1381

One other very important source of fourteenth century material is the Lay Poll Tax return of 1377 and to a lesser extent the returns of 1379 and 1381.

TABLE 3.1 THE 1327 EXCHEQUER LAY SUBSIDY FOR NEWTON ST LOE. TAXES LEVIED ON PERSONS HAVING GOODS OVER 10s IN VALUE*

J. de S.Laudo	7/-
H.atte Clegh	5/-
J.Wolmere	12d.
Thoma Rossel	5/-
W.atte Bergh	18d.
Willelmo Rayssh	2/6
Hen. le Shepherde	9d.
Hugone atte Wheole	4/-
Waltero le Frenshe	6/3
J.Bud	6d.
J.Pestors	6/-
J.Whitolf	9d.
J.Rossel	2/6
Willelmo le Smyth	3/6
Laurencio le Carpenter	12d.
T.Hayroun	4/-
Ja.de S.Laudo	7/6
H.Petit	4/-
Ja.Le Devenisshe	6d.
Ricardo Bruton	2/-

Summa XXe villate predicte
    LXVs iijd

*This is a literal transcription of the list and contains the characteristic inconsistencies of notation.

For estimating population the 1377 return is possibly the most useful since the tax was levied at the rate of one groat (4d.) per person for all lay inhabitants over 14 years of age, except for regular beggars, who paid nothing. Clergy paid one shilling each so it is a simple calculation to find out how many people actually paid tax in this year. The only problem is to decide what proportion of the total these taxpayers actually represent. Underenumeration for the counties with the least conscientious collectors is thought to be about 2½%. One authority suggests adding 50% to allow for those under the age of 14 and a further 5% to cover the exempt and the evaders.

TABLE 3.2 THE 1327 EXCHEQUER LAY SUBSIDY FOR THE SETTLEMENTS IN THE HUNDRED OF WELLOW

Modern name	Total tax £	s	d.	Total taxed	Av. per capita tax s.	d.
Woodborough and Peglinch	4	7	1	29	3	0
Tellisford and Farleigh Hungerford	1	4	8	13	1	10¾
Stony Littleton and Foxcote		15	3	6	2	6½
White Ox Mead		14	0	6	2	4
Whole of Wellow	7	1	0	54	2	7
Hinton Charterhouse		12	6	6	2	1
Norton St Philip		14	6	9	1	7½
Dunkerton		12	0	7	1	8½
Camerton		10	1	5	2	0
Combe Hay		18	9	10	1	10½
Corston	2	6	9	15	3	1½
Newton St Loe	3	5	3	20	3	3
Twerton	2	4	9	17	2	7¾
Englishcombe	incomplete			..	..	

Another argues for between 40 and 50% for the under 14s but up to 25% for undere-numeration. It is difficult for the layman to decide, when experts disagree, but another strategy that has been used in the past is to multiply the total taxpayers recorded by 1.5 to arrive at the total population, and at least this has the advantage of simplicity.

The main disadvantage of the 1377 returns is that for special reasons the inhabitants of the counties of Cheshire and of Durham were excluded.

The 1379 return was graded according to rank so one can get from it a rough idea of the social stratification of different communities. Again the returns for a particular settlement often gain in significance if they are compared with others in the district. A comparison of the 1377 Poll Tax returns with the 1334 or the 1327 Lay Subsidies gives an opportunity of estimating the impact of the Black Death in different districts.

3.2.4 Inquisiciones Post Mortem

The death of a feudal vassal brought to the king certain income. He had to be informed of the date of death, the age of the vassal's heir and the value of the holdings that were to pass to the king. These inquisitions contain valuable data for estimating the expectation of life in England between 1219 and 1509, but of course the sample is heavily biassed in favour of the wealthier and more privileged classes and so we have to be careful about applying the results of an analysis of their characteristics to the population as a whole. Many of the inquisitions are preserved in the Public Record Office and many have been published. Probably their main value for the historical geographer lies in the light they shed on the landholdings of the deceased, the amount of land held and the geographical spread of the manors in which it was held.

3.2.5 Ecclesiastical assessments 1254, 1291 and 1341

There were a number of assessments made by the church in the thirteenth and fourteenth centuries for the purpose of raising money. The three most important are:

1. The Taxation of Norwich, made in 1254, occasioned by a tax made on the English clergy by Pope Innocent IV at the instance of Henry III. Returns survive for five English and three Welsh dioceses and for certain religious houses. Extant returns have been published (Lunt, 1926).
2. In 1291 Pope Nicholas IV levied a tax granted to Edward I on taking the Cross. Although it has shortcomings it is the most comprehensive directory of medieval benefices in

existence and it is particularly important since it formed the basis of many subsequent taxes levied on the clergy.

3. In 1341 the Crown decided to take from the clergy a ninth of the corn, wool and lambs that came to them as tithes. Returns have survived for many parishes in the 27 English counties and were printed by the Record Commission in 1807 under the title *Nonarum Inquisiciones*. This assessment offers the opportunity of comparing the value of the parish in 1341 with the value in 1291.

3.3 OTHER OFFICIAL SOURCES

While there are very few sources that approach the Lay Subsidies for comprehensive national coverage there is a vast amount of material in the official documents of the state offices, kept in the Public Record Office. Many of these documents have been transcribed and published and can be consulted in large libraries. They are for our purposes most useful for the stray beams of light they cast on an otherwise dark period between the Norman Conquest and the Elizabethan period. If one is interested in a parish or a group of parishes it may very well be that an entry in one of these documents may give at least some insight into what was happening. Selective use of the very full indexes will make the task of searching them fairly undemanding provided that one resists the temptation to be sidetracked by interesting irrelevancies.

The following sources could be useful:

3.3.1. The Curia Regis Rolls

Printed from 1201 to 1237 inclusive, these are Rolls of the Central Courts in London and contain the record of important legal cases of many kinds. They are of especial interest to lawyers since they cover a critical formative period in the emergence of the English legal system.

3.3.2 The Pipe Rolls or The Great Rolls of the Pipe

These span the period from 1130 until 1241–1242 with some gaps. They contain entries relating to the revenue and expenditure undertaken by the powerful sheriffs of the counties on behalf of the Crown. The rolls above all are Exchequer documents, recording the official register of all the debts of the Crown answerable at the Exchequer. Like the Domesday Book the Great Roll of the Pipe was the unalterable evidence of the extent of the demesnes from which the annual revenue of the sovereign was mainly derived. A new edition of the Great Roll had to be drawn up every year.

An excellent description of the Pipe Rolls and the part they played in the financial system is provided by *Introduction to the Study of the Pipe Rolls*, published in 1884 by the Council of the Pipe Roll Society and reprinted by Kraus Reprint Ltd., Vaduz, 1966.

3.3.3 Calendar of The Close Rolls

Printed volumes span the period 1296 to 1509. The Close Rolls take their name from the Royal Letters Close, so called because being comparatively private and addressed to a particular person they were folded small and closed, apparently secured by a narrow strip of parchment having the Great Seal dependant and perhaps bearing the address. The copy in the Close Rolls was kept for reference. The Close Rolls of the thirteenth, fourteenth and fifteenth centuries contain entries of a most varied character, many of which are of value to the historical geographer.

3.3.4 The Patent Rolls

Printed volumes are available from 1216 until 1575. 'Patent' in this context means 'open' and is the opposite of 'close'. The patents contained in the Rolls relate to lands, grants, leases, licences to alienate, grants of wardship, special pardons and conveyances of land. An example of an entry under 'pardons' is: 'July 3 1432 at Westminster, William But, late parson of Newton Seintlowe, Co. Somerset, for not appearing before the same to answer Thomas Swynford of Bathe touching a plea of debt of 73s.'

3.3.5 The Fine Rolls (1272–1509)

These take their name from the enrolment on them of fines or payments made for writs, grants, licences, pardons and exemptions of various kinds, nearly all of them in effect bringing money into the Exchequer.

3.3.6 Rotuli Hundredorum, or The Hundred Rolls of the late thirteenth century

The most useful and best known of these are for 1279–1280. They represent part of a great survey recording the extent of the demesne lands, whether of arable, meadow, enclosed pasture or woodland, and of the holdings of the peasants together with their obligations and rents. The only counties for which the rolls have survived are Bedfordshire, Buckinghamshire, Cambridgeshire, Huntingdonshire, Leicestershire, Oxfordshire and Warwickshire, and translations exist for most of these in the local record society publications. Analysis of the returns suggests that even at this period there was a large amount of subleasing amongst the peasants as well as many examples of demesne land being worked more by wage labour than by peasants owing labour service. Subleasing allowed consolidation of scattered holdings and so promoted greater efficiency, leading in some areas to earlier and more effective enclosure. Peasants with small holdings who wished to do so could sometimes get rid of their land and work for cash, so stimulating the demand for goods and increasing the opportunities for trade and commerce.

3.3.7 'Pedes Finium' or 'Feet of Fines'

These must not be confused with the Fine Rolls. The word 'Fine' can mean either a fee or an end. In the sense used here it is applied to a final agreement or decision concerning landholding. From at least 1190 until abolished by act of parliament in 1834 the 'Feet of Fines' (*Figure 3.1*) was a near-perfect system of land registration. The agreement was originally written in triplicate on one piece of parchment, each party taking one part, the third part, written at or across the end of the other two and thus called the foot, being retained by the court as a record. Between this foot and the ends of the other parts was written in large letters the word 'Chiriographum', being divided by an indented cutting when the three were separated, part of each then being left on the foot. Hence, 'feet of fine'.

3.3.8 The Valor Ecclesiasticus

This was made in 1535 under the direction of Thomas Cromwell on the eve of the dissolution of the monasteries. It covers not only monastic land, but cathedral lands, bishops' lands, the land of hospitals and the glebe lands of the parish clergy, and shows for each the gross and the net income. A printed version was issued by the Record Commission in six volumes during 1825–1834.

When the church lands were in many cases sold by the Crown after the estates had been confiscated each parcel was carefully described. These descriptions are held in the Public Record Office as 'Particulars for Grants'. Some local and county record societies have calendared them and they should be consulted wherever possible. Not all of the original has survived. Details of Berkshire, Cambridgeshire, Essex, Hertfordshire, Northumberland, Rutland and some parts of Middlesex and Yorkshire are missing but a summary of all the incomes from the benefices in these places is complete.

Where information is available in such detail as in *Table 3.3* we have the opportunity to make a number of useful and interesting comparisons between the parishes in a particular area, and to compare the patterns that suggest themselves with those produced by studying other sources from other periods.

Figure 3.1 The form of the Feet of Fines

3.4 MISCELLANEOUS

3.4.1 Local record societies

Apart from the extensive series of records produced by official bodies in England there are many voluntary societies devoted to the publication of historical records of all kinds. *Table 3.4* gives some idea of the range of geographical coverage.

TABLE 3.4 LOCAL RECORD SOCIETIES (Davis, 1975)

Date of first publication	Society and (if necessary) district covered	Number of vols by 1974
1835	Surtees Soc. (between Humber, Forth Clyde and Mersea; based on Durham)	186
1845	Chetham Soc. (Lancashire and Cheshire)	245
1879	Lancashire and Cheshire Record Soc. (emphasis on relevant material from the P.R.O.)	114
1880	William Salt Soc. (Staffs.)	78
1885	Yorkshire Archaeological Soc. Record Series	133
1885	Oxford Historical Soc.	124
1887	Somerset Record Soc.	72
1891	Thoresby Soc. (Leeds)	51
1893	Worcestershire Historical Soc.	58
1902	Sussex Record Soc.	69
1903	Thoroton Soc. (Notts.) Record Series	30
1906	Devon and Cornwall Record Soc.	75
1911	Lincoln Record Soc.	68
1913	Bedfordshire Historical Record Soc.	53
1919	Oxfordshire Record Soc.	48
1921	Dugdale Soc. (Warwicks.)	30
1924	Northamptonshire Record Soc.	25
1930	Bristol Record Soc.	27
1931	Norfold Record Soc.	41
1938	Buckinghamshire Record Soc.	17
1939	Wiltshire Archaeological Soc. Records Branch	28
1952	Bristol and Gloucestershire Archaeological Soc. Records Branch	10
1958	Suffolk Record Soc.	16
1964	Dorset Record Soc.	5
1965	London Record Soc.	10
1965	Derbyshire Archaeological Soc. Records Series	4

TABLE 3.3 RECEIPTS OF THE CHURCH OF NORTH CERNY, GLOS. AS GIVEN IN THE VALOR ECCLESIASTICUS (after Bagley, 1965)

	£	s.	d.
In the parish:			
96 acres — arable land	1	4	0
70 acres — pasture		14	0
Tithes:			
corn	10	6	0
benefice and offerings	1	6	11½
hay	1	0	0
wool	6	0	0
lambs		17	0
cows and calves		8	1
flax, eggs, fruit, piglets and geese		5	10
Total	22	1	10½

Allowing for necessary payments of 11s.5d. the net annual value of the parish for tax purposes was £21 10s 5½d.

Apart from the record societies most counties have their archaeological and natural history societies, whose publications sometimes include volumes of 'Notes and Queries' that are another invaluable source of information.

Also well worth study is a small booklet by F.G. Emmison and W.J. Smith (1973) called **Material for Theses in Some Local Record Offices**. It contains a selection from lists made by the county archivists of material in their possession that they believe would repay further investigation. While the kind of research needed is more at the level of graduate dissertations in many cases there is still scope for original work of a more limited kind, involving much less time, on such subjects as river works; river trade; turnpike trusts; canals; slum clearance; the growth of public utilities like gas, water and electricity suppliers; gaols; estate agents' records; and auctioneers' accounts to name but a few. Many of these of course fall outside the scope of this chapter but this is an appropriate place to mention them.

As a general working principle it is sensible always to make the most extensive enquiries first at the local reference library. Further information may well be forthcoming from the local county record office though it is likely that original documents from the medieval period would be difficult for the uninitiated to decipher without some expert assistance.

The following examples relate once more to the village of Newton St Loe and show how useful even quite specialised sources can be in filling in important details.

First is a record of a deed drawn up on 2 February 1344 by John of St Loe and the Hospital of St John the Baptist in Bath. This deed has recently been published along with many others from the Hospital (Kemp and Shorrocks, 1974). It reads:

Concession by John de St. Loe (**Sancto Laudo**) knight, lord of Newton St. Loe (**Nyweton'**) by Bath, for the safety of his soul and that of Elia his wife lately dead, that, whereas Sir Roger de St. Loe (**Sancto Laudo**), lord of Newton St. Loe (**Nyweton'**), before the Statute of Mortmain granted in free alms to the master and brothers of the Hospital of St. John Baptist, Bath, **the meadow lying in the great meadow called Brodemede, between the plot of cultivated land (*cultura*) called Herddelond within the demesne of Newton St Loe (*Nyweton'*) on the east and the said meadow on the west,** to be held quit of all secular extraction; they may hold it in free alms for ever as above said, quit of any payment or custom.

Sealing witnesses: John of Petherton (**Pederthon'**), John in the Clay (**in la Claye**), Henry Parvus (**Petit**), Richard le Brutton', William Wardde, and others. Given at Newton St Loe (**Nyweton'**), Wednesday, the feast of the Purification of the B.V.M., 18 Edward III.

Seal Medium-sized, round, red was, showing family arms upon a triangular shield, viz., a belt or band to the right, with, on the upper part of the shield, a transverse line with five hanging tags (i.e., a bend dexter with a label of five points in chief). Inscription: *SIGILLVM: IOHANNIS . . . (the rest broken).

Note that two at least of the signatories bear the same family name as on the 1327 Exchequer Lay Subsidy return transcribed in *Table 3.1*.

The second example, although straying a little outside the scope of this chapter, is included because it relates to the same theme as the first extract. It is from a 'Survey and Rental of the Chantries, Colleges, Free Chapels, Guilds, Fraternities, Lamps, Lights and Obits in the County of Somerset' undertaken in 1548, which was published by the Somerset Record Society in 1888. Under Newton St Loe it is recorded:

Land and possession given to the use and maintenance of a lamp and a light burning in the parish church there....

John Grevys, clerk, rector of the parish church there *holds 4 acres of arable land lying in the north field there*, and renders per annum 2s. 8d.

The same holds *3 acres of arable land lying in the south field* there and renders 16d. Total 4 shillings .... It is presented that this rent groweth but every second year.

The two extracts suggest that by 1344 there were already separately identifiable parcels of land in the manor that may either have been part of the common fields or have represented early enclosures of waste or common land. By 1548 the reference to the north and to the south field might lead one to infer that the two rather than the three field system may still have been in operation, but there is another Bath Hospital charter dated early in the thirteenth century that refers to land in the east field and it is not possible to say whether this is evidence of a third open field, the name of an early enclosure, or whether it opens the possibility of a fourth, west field that has not yet come to light. It is only by assembling every scrap of evidence that one is able to draw up a balance of probabilities. Sometimes it is possible to work backwards from a later document to clear up difficulties in an earlier period.

REFERENCES

Bagley, J.J.(1965). *Historical Interpretation.* Harmondsworth, Middx.; Penguin

Davis, R.H.C. (1975). 'Record Societies in England.' *History, 60*, 239

Emmison, F.G. and Smith, W.G. (1973). *'Material for Theses in Some Local Record Offices.'* Chichester, Sussex; Phillimore

Glasscock, R.E. (Ed.) (1975). *The Lay Subsidy of 1334.* London; Oxford University Press for the British Academy

Government Publications, Sectional List No. 24. *Record Publications.* Revised January 1977. HMSO. This includes a complete list of all the calendars available of the official sources such as the Patent Rolls etc., some of which have been listed and described in section 3.3

Kemp, B.R. and Shorrocks, D.M.M. (Eds) (1974). *Medieval Deeds of Bath and District.* Somerset Record Society, Vol. 73

Lunt, W.C. (1926). *The Valuation of Norwich*

Martin, C.T. (1976). *The Record Interpreter* (Abbreviations, Latin words used in English historical manuscripts and records). Reprint. Kohler & Coombes

Mullins, E.L.C. (1958). *Texts and Calendars, an Analytical Guide.* London; Royal Historical Society

CHAPTER 4 AGRICULTURE

4.1 INTRODUCTION

This chapter covers a period from after Domesday until the end of the nineteenth century.

It is in no sense intended to be a short history of British farming, but an exploration of some of the many sources of information about farming conditions, practices and characteristics mainly at a local scale.

Some very brief introductory comments are made to provide some sort of context against which particular sources can be seen. For example, the interest in improved farming practice that became apparent towards the end of the eighteenth century had a great deal to do with the economic and social conditions of the time, and is difficult to understand without at least some reference to wider issues.

It will always be necessary for the student tackling a particular place and period to read widely, to get a perspective on the broad issues and processes at work and to look carefully at case studies made elsewhere for the same period and at a comparable scale.

4.2 THE MEDIEVAL RURAL ECONOMY

After the detailed picture presented by the Domesday Survey it is centuries before an equally illuminating documentary source is available covering the whole country, so we have to build up a picture of some of the most important changes from individual studies and from scattered scraps of information for the most part. There is still room for a great deal of disagreement between scholars and many of the ideas that were held about the nature and causes of change in the past are now being challenged to some extent and it seems as if modern scholarship has increased uncertainty about some aspects of the past. Certainly there are many gaps in our knowledge. It has been said that in this period we know too much about peasants and too little about landlords. This is an overstatement. We certainly know too little about the smaller landlords whose estate documents, if they ever existed, have not survived like those of the great monastic houses or the important lay estates. We know almost nothing about the peasants whose landlords were knights of the shire or even franklins of the village, and yet it was often in these areas, freed from the uniformity of the large landowners, that one would expect to find great variety in farming organisation and practices. So it is sometimes difficult for the non-historian to keep his bearings in a changing landscape.

Two of the standard works on the manorial economy, H.L. Gray's *English Field Systems* and C.S. and C.S. Orwins' *Open Fields* are half a century old now. Gray explained the common field system as a scheme of cultivation brought ready-made over to England from the Continent by the Anglo-Saxons. The Orwins saw it as a commonsense response by a pioneering society to conditions in which the best way of minimising hardship or disaster was some form of co-operative organisation. Over the past 20 years evidence has come to light that makes some scholars, at least, unhappy with the traditional views of the evolution of the open or common field system. The debate is sometimes technical and difficult for the layman to follow. It has been well set out in a recent collection of articles on social history (Hilton, 1976) particularly in the dialogue between Dr Joan Thirsk, who challenges the traditional view and Dr J.Z. Titow, who defends it.

There are certain features of the common field system however about which there is general agreement. First, the arable and the meadow land were divided into strips among the cultivators, each of whom could occupy a number of strips scattered about the great fields. Second, both the arable and the meadow land were thrown open for pasturing by the stock of all the commoners after the harvest had been gathered in, and in the seasons when a particular piece of the common land had been left fallow. Third, there was common pasture and waste, where the cultivators of the strips had the right to graze stock, to gather timber or to cut peat. Fourth, the decisions about who did what, and when, were made by an assembly of the cultivators, who generally formed the manorial court in most places in the Middle Ages. Many of these courts lasted into the eighteenth century in some districts.

Dr Thirsk argues that the oldest element in the system is probably the rights of common grazing over pasture and waste, and that such rights probably existed long before even the Saxons came to England. The other components, she believes, need not all have been present in the village community from the beginning of settlement. There are some villages lacking one or other of the components as late as the later Middle Ages:

Yet from the sixteenth century onwards, manorial documents contain more and more explicit details about the working of the system until in the seventeenth and eighteenth century they are at their most emphatic and lucid. On the eve of Parliamentary enclosure some maps of common field villages present a more orderly pattern of strips, furlongs and fields than anything available earlier.

Here, argues Dr Thirsk, are the grounds for believing that the system evolved slowly. She

Figure 4.1 The main area (⟋⟋⟋⟋⟋⟋) of open field agriculture in England (based on Gray, 1915)

thinks that the common fields operated by the village communities evolved, in England, in the twelfth and the first half of the thirteenth century.

Dr Titow on the other hand vigorously defends the more traditional view that would place the existence of the fully developed open field system back as far as the tenth century.

The best simple indication of the areas of England in which the common field system was found is that given by the Orwins and based on the work of Gray (*Figure 4.1*) It is in these areas that the village community seems to have had that high degree of coherence and organisation which allowed it to survive intact in many cases through Tudor and Stuart times and well into the Georgian period. It is in these areas that the enclosure of the common fields usually had to be effected by Act of Parliament.

Outside these more central areas the open field system was on the whole much less well developed and the process of enclosure often took the form of a gradual encroachment on the common pasture land or on the waste.

In many of the peripheral areas stock were at least as important in the manorial economy as arable land, and this made the process of change much more natural. For the mainly pastoral villages, interest in strip-cropped common fields was slight and for them enclosure was usually a painless, peaceful and slow process. The 'early enclosed' areas of England, central Suffolk, most of Essex, Hertfordshire, parts of Shropshire, Herefordshire, Somerset, Devon and Cornwall were mostly pastoral districts in the sixteenth century and their enclosure was not a difficult process because they had never experienced the kind of highly developed open field system such as was found in the Midlands.

It is likely that the earliest of the open field communities, by a process of trial and error, hit on the expedient of dividing the cultivable land into two halves, to be tilled in alternate years. As time passed and the population grew it may have become necessary to till two-thirds or even three-quarters of the land each year, leaving only a small portion fallow. Under such circumstances the land would soon lose its fertility, and it would be seen that a fallow period of one year in two, or at the least, of one year in three, would be needed to allow the land to recover. It may well be that the two field system gradually evolved into the three field system in such a manner. Normally it would not have been necessary to go as far as to have a four field system, at least until the Norfolk four course rotation was widely adopted in the eighteenth century, by which time enclosure of the open fields was seen as the means to greater agricultural efficiency.

Other changes took place in the status of the feudal classes. In particular Hilton (1976) has argued that the inflation of the late twelfth century caused a catastrophic decline in the status of the villeins between 1180 and 1200. The decline in the value of money forced the landlords to respond by attempting to stem the fall in the value of the cash income they got from their tenants by requiring them to provide labour services rather than cash. There is evidence that since Domesday there had been a growing tendency on the part of the villeins to commute labour service for cash payments. Landlords could then use this cash to hire labour. With the inflationary situation it is believed that landlords fell back on their ancient right to demand labour service and so the villeins as a class and labour service with it persisted far longer in England than it otherwise would have.

The Curia Regis Rolls (*see* page 35) contain many references to cases recording a head-on clash between lord and tenant, the one insisting that the other is a villein and unfree, the other asserting his freedom. In many of the cases the tenant was the plaintiff and it suggests a very strong sense of injustice for the inferior person to take the initiative in a Royal Court. One reason for the tenant wanting to prove his freedom was that if he were so recognised he would be free to continue to buy and sell land as well as to avoid irksome personal service. The lay subsidies and other documents such as deeds and land grants show that by this time there were numbers of people in many communities who were far from being paupers; parcels of land in and around the open fields were leased, owned, sold or given in alms in increasing numbers, and new fields and enclosures had been created by individuals from land previously uncultivated, whether waste, common land or woodland.

It is now clear that the half century from 1200 to the Black Death was a disturbed period in England. The causes of the inflation were complex. It has been suggested that there was a Malthusian crisis as a result of an earlier growth of population and an extension of cultivation into marginal land which suffered quickly from too intensive cultivation and gave rise to localised famines. There is evidence of extensive decimation of cattle and sheep from disease. The wool industry went through a difficult period, arable cultivation contracted and agricultural production fell, and many of the population suffered increasing distress as well as a reduction in status. The Black Death came not so much as a bolt from the blue but as the climax to a series of difficulties.

The substantial fall in population caused by the Black Death changed the situation. The bargaining position of the peasants was improved and reductions in rent and in demands for service were partly the result of stiffening peasant resistance to the lords as they recognised the power of their position. Through the later fourteenth and fifteenth century many tenants improved their position, accumulating landholdings, consolidating and even enclosing their lands in some cases. On the other hand the lot of many of the peasants declined to the point where they became landless labourers dependent upon others for employment and support.

4.3 MANORIAL COURT RECORDS

The manorial courts that were established in the early Middle Ages settled such matters as the choice of officials and the pattern of husbandry for the coming year, set fines for petty offences, recorded the admission of new tenants in the manor and set out what services they had to perform. The courts continued in many places well into the eighteenth century. The court rolls of Newton will be quoted to show the value of such documents in reconstructing the geography of a parish in the past. Most fortunately at the Court Baron of the Lord of the Manor, held at Newton on 25 October 1672 the village representatives met to acquaint their lord with all the ancient customs, privileges and dues associated with the copyholders (*see* Glossary) of the manor.

The next reversioner or the lord may enter into all the ancient meads and into all the arable land not enclosed within the quickset (i.e. a living hedge) or other durable bounds lying in the field called the Fallow Field. The residue of the tenement (i.e. namely all the houses, courts, gardens, bartons, orchards, closes of pasture or arable and all the arable land lying in common in the field then called the Common field) the executor is to hold... till the Feast Day of St. Michael....

Then later it was stated that no copyholder could let his meadow or pasture for more than a year without permission from the lord of the manor. Nor could he 'convert any of his meadow into arable, or fell any of his timber, to burn, set, root up or neglect his bounds and lay his ground common, spoil his fruit trees, suffer his house to be ruinous, refuse to pay his rent...'

Finally his tenants told their lord that there was a plot of land in Newton Mead that belonged to the lord's demesne. It was divided into two parts known as the First Mowth and the Latter Mowth, upon which the tenants were accustomed to cut hay. The cutting operation was organised as follows: The tenants arranged the date for cutting amongst themselves. It was usually in the last few days of June. Twenty-three mowers and 13 haymakers were sent to spread and turn the grass and to make it into cocks, first in the First Mowth and then in the Latter Mowth. The lord provided four haymakers at his own expense the first time, and five the second. Four specified tenants had to pay the mowers the market price on 11 November of 'six bushels out of right wheat'. The lord allowed each of the mowers to take away for his own use the grass from a parcel of land called a Judd measuring six feet by 99 feet. Fourteen wagon-loads of hay had to be taken to the manor house before the tenants collected their own hay and for the carting the lord provided 'a sufficient portion of bread, cheese and beer'.

The court rolls are also informative about other aspects of farming life and practice. Evel's was the name of the village pasture and there are many references to it in the rolls. The following extract was written in 1720.

IMPRIMIS we present James Davis for a new Hayward with the keeping of Evel's Bounds and all the customs and privileges belonging thereto.
ITEM we do order all persons that have bounds against the common mead to repair the same before All Hallowtide with the cleansing of the ditch at the same time and that the ditches be duly cleaned once in three years upon penalty of five shillings for every person making default.

The ditches referred to are those on the inside of the quickset and were used by cattle for shelter in cold or in wet weather. There was also a reference to 'overpressing' (i.e. overstocking) on the commons and everyone defaulting was ordered to pay the five shilling fine to the driver before removing the offending cattle.

4.4 THE ENCLOSURE MOVEMENT

Enclosure of the common fields in England was a process that had been going on for centuries. In the period after the Norman Conquest it is clear that there were two kinds of enclosure. The first was the enclosure of the wasteland that on many manors lay beyond the open fields. In 1235 and 1285 the Statutes of Merton and of Westminster II gave power to the feudal lords of the manor to enclose uncultivated waste not needed by their free tenants, though on the whole not much land was enclosed under these statutes. The second was the enclosure of land from the common fields, usually a slow and piecemeal process, more evident in the areas of pastoral and mixed farming than in the heart of the Midlands where the three field system of open fields was firmly established.

From the late fifteenth century till the beginning of the seventeenth century there was an endless flood of propaganda alleging with some justification that enclosure by landowners to increase pasture for their sheep, cattle and deer was at the expense of arable land for the support of the rural population and so was responsible for the depopulation of rural England. Sir Thomas More's *Utopia*, published in 1516, claims that the agrarian revolution then under way was damaging the interests of the small proprietors. Although the government made periodic attempts to check the process they were not very successful. By the seventeenth century enclosure was effected by the Chancery or Court of the Exchequer and in these enclosure awards it was not unusual for the peasant interests to be taken (if inadequately) into account by making special allotments of land for the poor. Many thousands of acres were enclosed by these means, especially in Durham, Lancashire and Cheshire.

From about 1760 the normal means of effecting enclosure came to be by private act of Parliament. The promoters of each act nominated in it the commissioners who visited the parish, heard all the claims of those having either open land or common right and then they allotted to all those who had made good their claim an equivalent amount of land, which was then normally fenced or hedged to produce the distinctive landscape over much of England of small enclosed fields and scattered farmhouses.

The decisions arrived at were embodied in the Enclosure Award (*Figure 4.2*) which Tate (1967) has described as 'the foundation charter of the modern village'. The earliest enclosure

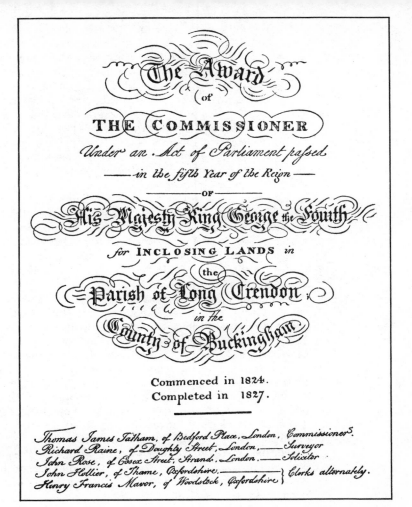

Figure 4.2 A typical Enclosure Award, Long Crendon, Bucks., 1827 (enclosed 1824—27) (From Tate, *The English Village Community*, 1967, p. 116, courtesy of Victor Gollancz Ltd)

act in England was probably that for Radipole in Dorset, passed in 1603. After 1760 there were about 5400 enclosure acts, covering more than seven million acres, about one-fifth of England. Like the Tithe Award maps the maps made for the Enclosure Award are invaluable sources of detailed information ranging from field names to drainage. Most of the awards and the maps are still available and usually can be consulted in the county record offices.

Historians writing about the process of enclosure have by no means shared a common view about its effects. One of the most readable sources even today is J.L. and B. Hammonds' **The Village Labourer** which is passionately biassed in favour of the peasantry, and strongly critical of all the other classes in society that were involved in the movement. It has been described by one acid critic as one of the most brilliant pieces of historical fiction in the English tongue. Other writers take equally extreme positions in the other direction. Modern scholarship, with more information and further in time from the events it is describing, tends to take a more dispassionate view, and there is some support for the 'middle' position, namely that the social costs of enclosure were considerable, often greater than were necessary, disadvantaging the small proprietor, but that the technical advantages brought about by the incentives given to the farmers to improve their land by new methods broadly justified the changes. But the last word has by no means been written on the subject and it is perhaps wisest to follow Tate, who has worked on the theme longer than most, and be prepared to keep an open mind.

Where the enclosure was a gradual process and there was no parliamentary award then it can be a long, difficult and sometimes impossible task to establish how and when its various stages were accomplished. We have already discussed the variety of sources of information about the different parcels and lots of land that were the source of charitable bequests, inheritance or legal disputes over ownership from the early Middle Ages onwards.

When one is studying a particular parish an assiduous search will usually produce a large number of references to such pieces of land but never is there anything remotely resembling a map until the seventeenth, or more usually, the eighteenth century. It is not uncommon to find maps, either in estate offices, the parish chest or the local record office, on which the names of the fields are recorded. The field names themselves can give clues to when and for what purpose the fields were created.

The map in **Figure 4.3** shows that by 1742, even if a certain amount of artistic licence had been taken by the cartographer, the land in the parish was already largely enclosed, with Newton Park, around the mansion of that name, and Newton Mead, fringing the river in the north, being the main still open areas.

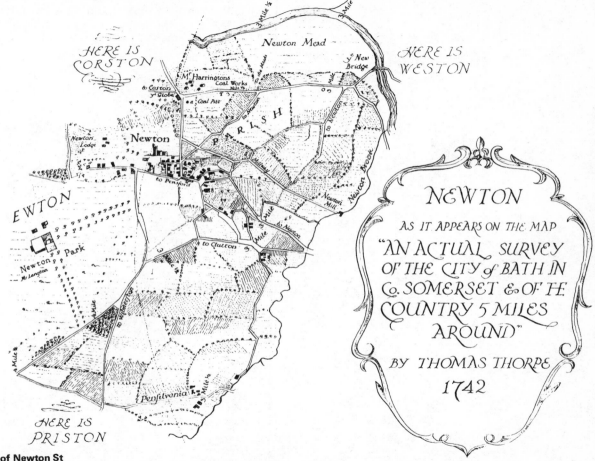

Figure 4.3 The Village of Newton St Loe as it appears on a map by Thomas Thorpe, 1742

Figure 4.4 Field names in Newton St Loe

A slightly later estate map of the parish (*Figure* 4.4) names many of the fields. 'Tyning' is a common name and means newly enclosed land. Personal names followed by 'tyning' can, with a diligent search of the parish registers and rate books, sometimes allow one to say that a certain field could not have been enclosed before a certain date. 'Evels' we know from the manorial court records to have been common pasture in 1692. 'Glebe' indicates land belonging to the church and it is found both in the north and in the south of the parish, possibly indicating a relic of an earlier two field common field system, as was hinted at in the extract from the 1548 Survey of Somerset Chantries (*see* page 40).

Of the other names, many indicate either the use, the size or some other characteristic of the land. 'Leaze' means meadow. 'Ham' usually refers to low-lying damp pasture land beside streams or rivers. 'Cleeve' means steeply sloping land (hence cliff). 'Rudge' means partridge and usually indicates a covert of some kind. 'Conygre' comes from 'coney' (rabbit) and means land with rabbit warrens upon it. Rabbits were a prized source of meat and were introduced into England by the Normans; they flourished on sandy soils. The term 'ball' was used frequently from the fifteenth to the seventeenth centuries to denote animals, so the 'ball acres' were probably used for pasture.

However caution is needed in interpreting some of these old names: many apparently obvious derivations turn out on investigation to mean something quite different. The use of a good place or field name dictionary is recommended where there is any doubt at all about the meaning of a particular name (*see* References).

4.5 TITHES AND TITHE COMMUTATION

In the eighteenth century the influence of the State over the Church in England had never perhaps been greater. Privileged clergymen formed part of a network of patronage and influence that effectively dictated how the country was ruled. The bishops, even the poor ones, like the one at Bristol with a mid-century income of less than £500 per annum (compared with £7000 for Canterbury), were on the whole comfortably off because in addition they could hold a number of rich deaneries and rectories 'in plurality' (i.e. simultaneously). There were other wealthy livings, parishes endowed with large incomes, but men were appointed to these for favours given or expected or for their family connections, and rarely if at all for their piety or pastoral competence and concern. Many of the livings brought in very small incomes and the poorest clergy, the curates, were on occasion forced to seek work outside the church, or to undertake commitments in several parishes, in a less lucrative form of plurality, and were often looked down upon by the small farmers of the country districts.

The value of most livings was dependant on land, and tithes were pre-eminently linked with land. The technical classification of tithes is into Predial (arising from the produce of land), Mixed (coming from the stock on land) and Personal (arising from the industry of the occupiers), but the alternative classification into 'great' and 'small' is more convenient. Great tithes were usually corn, hay and wool; small tithes wood, livestock and garden produce.

Vicars on the whole tended to be endowed with small rather than great tithes but the divisions were by no means unalterable. Curates were usually supported by small grants of land which the incumbent could either farm himself or lease to someone else. The income of rectors, vicars and curates in the end depended on tithes from the produce of the land.

The tithe, the tenth part of the produce of the land, was the traditional means of support for men of God, not only for many centuries in England and Europe but in many other societies. But by the eighteenth century, English society had moved a long way from the subsistence economy of previous centuries and tithe payments were increasingly irrelevant, it seemed, to many sections of the community. The widespread unpopularity of the tithe was in part due to the fact that it fluctuated widely in its yield as the harvests varied, partly because extremely complex legislation had grown up around it, and most directly because tithing in kind (its most basic form) appeared wholly inappropriate in the middle of the eighteenth century and was being superseded by money payments in some areas. Another powerful objection to it was that it bore exclusively on the farming interests. At this time new fortunes, founded on commerce and industry, escaped the compulsory tithe whereas farmers and other agricultural interests could not; indeed where enterprise and farming skill were employed to improve farming it was a source of frustration that the tithe owners extracted their dues even though they had in no way contributed to the heavy capital expenditure that produced the extra yield. It was in the period after the 1770s that the challenge to the tithe system reached its greatest intensity.

The Parliamentary enclosure movement in the eighteenth century created much of the present English landscape, the neat regular fields and scattered farmsteads being carved out of the great open fields that had persisted since the early Middle Ages. Enclosures usually made agricultural improvement easier, giving to landowners rents that could be doubled or even trebled within a short space of time. When the enclosure award was made it was not at all uncommon for tithes at the same time to be commuted, either by the allotment of land in lieu, or by cash payments. Most of the eighteenth century open field enclosures were in the English Midlands and open field enclosures were about twice as common as the enclosures of commons and waste lands, in which commutation of tithes was less common at this time. When the matter of commutation was being discussed one of the most important questions to be asked was how much land should be given instead of the tithe?. Tithes yielded different amounts each year as the harvest fluctuated and naturally the tithe owners wanted an arrangement that would give them the best terms. On the whole the advantage lay with the tithe owners who were able to drive hard bargains because if they dragged their feet, the people who wanted to enclose to increase their incomes would be prevented from starting these profitable activities. Tithe owners also had an enclosure commissioner to look after their interests and they were frequently allotted the land of their choice in prime sites.

The first task was to establish the boundaries of every district in which tithes were paid separately. This was known as a tithe district to distinguish it from a parish. Next the total value of tithes payable in each district for the previous seven years was computed from the actual receipts of the tithe owners and opportunity was given for the farmers and tithe owners to agree on a valuation by I October 1838. Where agreement was not reached the commissioners imposed the award.

Then it was necessary to apportion the rent charge amongst the lands in the district, a difficult matter because quality, fertility and value of land varied widely. In practice the only fair way to allocate dues was by reference to the observed state of cultivation.

Therefore very detailed surveys were made in many cases, the state of cultivation being entered as 'arable', 'grass', 'meadow' or 'pasture', 'common', 'wood', 'coppice', 'plantation', 'orchard', 'hop ground' or 'market garden'. In general the most important distinction was between arable land, regularly cropped and ploughed, whose tithes amounted to about one-fifth of the value of the rent, and permanent grassland whose tithes represented less than one-eighth of the rent. Almost all of the 11 000 surveys in England and Wales were made before 1851, the majority before 1841.

Details of the survey are set out in two documents, a map and an apportionment. Assistant Commissioner Dawson late in 1836 suggested using the opportunity presented by the Act to make a general cadastral survey of the whole country which could be kept down to a cost of 9d. per acre, but Parliament and the landed interests would have none of it. As a result only about a sixth of the maps used in the awards were specially surveyed for the purpose and of a sufficiently high standard of detail and accuracy to be given the Commissioners' seal and to be cited as a true legal record, if necessary in the courts. The

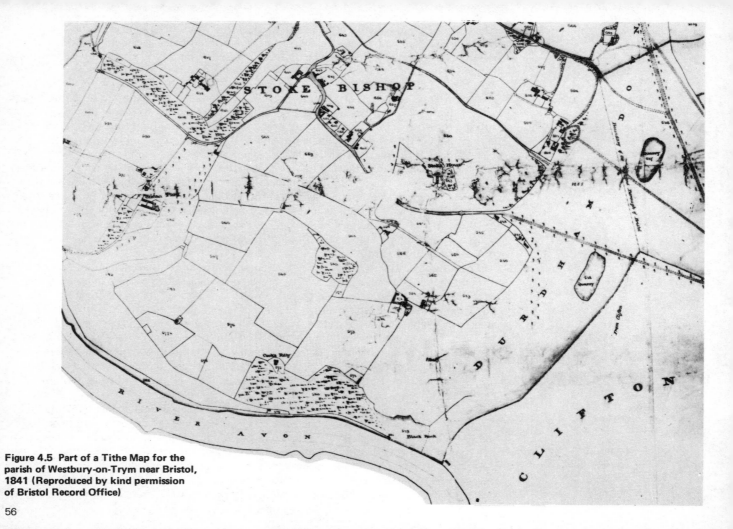

Figure 4.5 Part of a Tithe Map for the
parish of Westbury-on-Trym near Bristol,
1841 (Reproduced by kind permission
of Bristol Record Office)

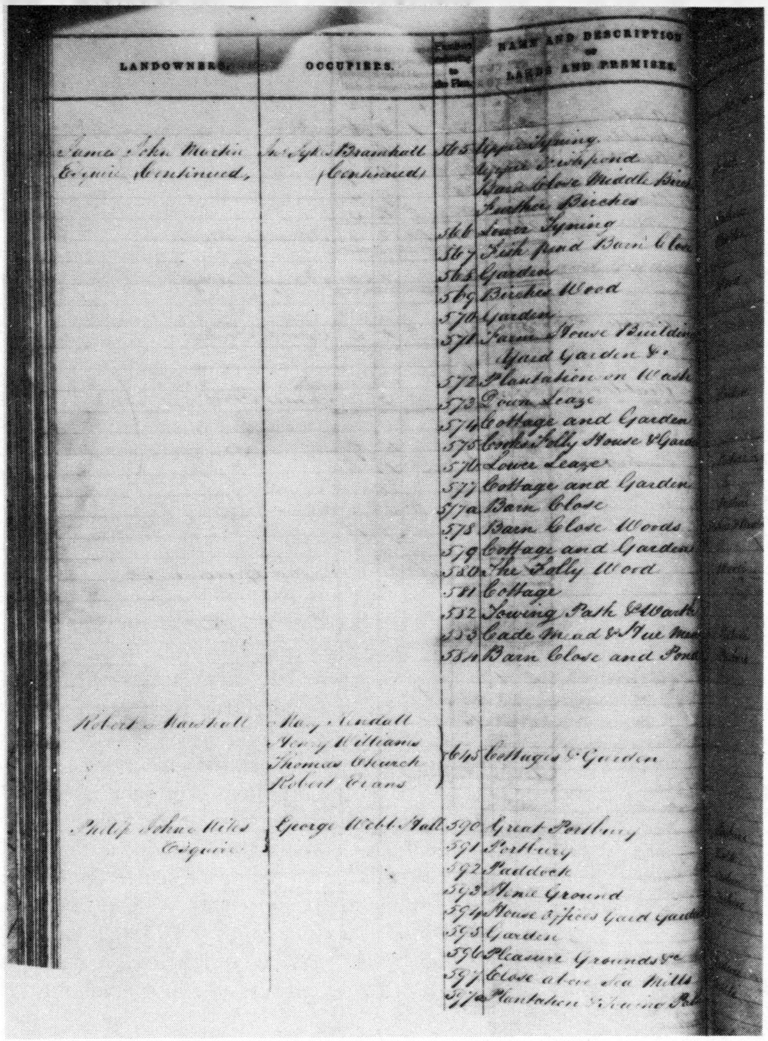
LANDOWNERS.	OCCUPIERS.	Numbers referring to the Plan.	NAME AND DESCRIPTION OF LANDS AND PREMISES.
James John Martin Esquire Continued,	Joseph Bramhall (Continued)	565	Upper Tyning
			Upper Fishpond
			Barn Close Middle Birch
			Further Birches
		566	Lower Tyning
		567	Fish pond Barn Close
		568	Garden
		569	Birches Wood
		570	Garden
		571	Farm House Buildings Yard Garden &c
		572	Plantation on Waste
		573	Down Leaze
		574	Cottage and Garden
		575	Cooks Folly House & Garden
		576	Lower Leaze
		577	Cottage and Garden
		577a	Barn Close
		578	Barn Close Woods
		579	Cottage and Garden
		580	The Folly Wood
		581	Cottage
		582	Towing Path & Wharf
		583	Cade Mead & New Mead
		584	Barn Close and Pond
Robert Marshall	Mary Kendall Henry Williams Thomas Church Robert Evans	645	Cottages & Garden
Philip John Miles Esquire	George Webb Hall	590	Great Portbury
		591	Portbury
		592	Paddock
		593	Stone Ground
		594	House Offices Yard Garden
		595	Garden
		596	Pleasure Grounds &c
		597	Close above Sea Mills
		597a	Plantation & Towing Path

Figure 4.6 Part of an entry in the Schedule of Apportionment for the parish of Westbury-on-Trym (Reproduced by kind permission of Bristol Record Office)

scale of these official maps was either three chains to an inch (approximately 26.7 inches to the mile) or six chains to an inch (approximately 13.3 inches to the mile). Some maps of large parishes with a number of detached plots measure as much as 100 square feet. For the rest the Commissioners were authorised to accept local surveys, as long as none of the parties concerned objected, even if these were known to be inaccurate. Though some are inaccurate and though most are less accurate than the official maps, many are still of a high standard (*Figure 4.5*).

The Apportionment is a roll of parchment sheets, officially required to measure 21½in x 18¾in, consisting of three sections of which the second is the most important. This is the Schedule of Apportionment in which each tithe area, numbered in the accompanying plan, is listed under the name of its occupier and its owner (*Figure 4.6*). Where the Act was being applied to a parish that still lay in open fields as many as 3000 tithe areas may be enumerated. Where, as was usually the case, a tithe area was a field, the field name is recorded; where it is not a field it is described as a 'piece of water', 'a chalk pit' or 'a house and garden'. The state of cultivation is entered, and the value of the rent charge apportioned to it are stated.

Three statutory copies of the map and the Apportionment were prepared. One is now in the custody of the Tithe Redemption Commission in Finsbury Square, London, EC2. A second copy was originally deposited with the incumbent and the Churchwardens to be kept in the parish chest. A third copy was kept in the diocesan registry. The second and third copies unhappily are not always where they were supposed to be. If not damaged beyond recall or mislaid, they are sometimes to be found in the local county record offices.

The many thousands of field surveys that were made under the Act depict with remarkable clarity and fine detail the rural landscape of England and Wales exactly as it was in the 1840s.

The maps show for each district the boundaries of woods and fields, roads and streams and the position of all the buildings. The schedules in the Apportionment give the names of their owners and occupiers, the area of the fields and plots of land and their state of cultivation. For the geographer they provide a mine of information. They contain examples of every type of landholding, from the scattered plots in the remaining open fields to the great estates of the landed proprietors. We can find from them how much titheable land belonged to estates of various kinds, how much land was owned by the universities, the Church and later by the railway companies. We can find out how much land remained unenclosed in 1840, where it was and who owned it. We can find out what land was farmed by owner-occupiers and what by tenants. The proportion of land under different cropping patterns is also given and can be usefully compared with present land uses and related to the underlying geology or to the soil if we are fortunate enough to find a suitable map.

Other information on tithes may be found in the parish chest. Here we sometimes come across cases submitted for legal opinion by a new incumbent anxious to make the most of the tithes to which he was entitled, and in investigating it the counsel may well have unearthed much useful detail about a particular parish.

When we consider the tithe surveys in conjunction with the 1841 census, the enumerators' books, private estate accounts and the splendid county reports in the form of 'Prize Essays' published in the *Journal of the Royal Agricultural Society* between 1845 and 1869 we have all the ingredients to present a lively and fully detailed analysis of rural England in mid-century.

4.6 AGRICULTURAL STATISTICS

4.6.1 Before 1866

In the course of the second half of the eighteenth century Great Britain virtually ceased to be self-sufficient in grain. In the last 12 years of the century the situation in agriculture grew serious. Population growth increased the demand from the home market and poor harvests in 1789 and 1790 kept prices high. Farmers and landowners were able to persuade the government to give them protection against a flood of cheap corn from abroad despite opposition from the trade and industrial interests. The war with France and more bad harvests aggravated the situation and increased social unrest to dangerous levels.

There were many who had long argued for the government to collect and publish accurate statistical information about crop production, yields and prices and after bad harvests earlier in 1756 and 1757 limited reports were made by individuals on their own initiative. Conditions by 1800 had reached the point where the government, alarmed by social unrest and the prospect of yet another bad harvest allowed the Home Office through the clergy to make a survey of conditions in each parish. When the Board of Agriculture was asked to comment on the completed returns they said they were 'so extremely erroneous that they cannot safely be at all relied on...' Modern research suggests a kinder verdict though it is generally agreed that the reports err on the side of understatement as far as acreages are concerned but they are useful for general comments and for details they give of the practices of husbandry at this time.

The wars with France brought out a sense of solidarity amongst the farming interest and country gentlemen, and the really large landowning classes began to draw together to further their own particular interest. In 1793 the Board of Agriculture was set up by the government, some of its tasks being to make a survey of farming in the country as a whole, to act as a forum for new ideas and to press for a bill to cheapen the process of enclosure. The Board was by no means an unqualified success. The members of the Board represented an aristocratic oligarchy and it failed to win the confidence and support of the gentry and farmers. The general survey was a poor substitute for a proper agricultural census and the Board only achieved this particular goal by commissioning a series of county reports under the general heading of 'A General View of the Agriculture of....', followed by the name of the county concerned. There are usually two reports for each county, one published within the period 1793—1796 and a second edition often by another writer between 1805 and 1817. The 'General Views' were rather subjective accounts, laying stress more on 'Improved Farming' than on the average state of farming in each county. On the other hand there is very little else written on this scale in anything like a systematic fashion in this period and so the volumes are an important source, to be consulted, but always with reservations about their accuracy and bias.

Arthur Young wrote the report on Oxfordshire. His method of collecting material has been criticised as haphazard and it is hard to tell where the truth actually lay between his enthusiasm for improvements effected and his impatience with what he considered backwardness. He comments trenchantly on the gulf between the progressive farmers with enclosed fields and forward-looking attitudes and the conservatism of the old type of open field farmer.

When I passed from the conversation of the farmers I was recommended to call on, to that of men whom chance threw in my way, I seemed to have lost a century in time, or to have moved a thousand miles in a day. Liberal communication, the result of enlarged ideas, was contrasted with a dark ignorance under the covert of wise suspicion; a sullen reserve lest landlords should be rendered too knowing; and false information given under the hope that it might deceive......The old open-field school must die off before new ideas can become generally rooted. (Young, 1809)

The Board of Agriculture was a disappointment. It had little money to encourage technical change; the county reports were not well written and sold poorly; and the General

Enclosure Bill was not very effective, emasculated as it was by the combined opposition of the Church and the Law. The Board ceased operation in 1821. Interest in the collection of agricultural statistics was low in the 1820s and 1830s but by 1844 the matter was raised for the first time in Parliament. Mr Gladstone pointed out that the main difficulty was to find a means of setting up appropriate machinery to do the task. Localised attempts were made in 1844–1845 and an effort to introduce a bill for the systematic collection of statistics in England and Wales was defeated in 1847 because the Chancellor of the Exchequer would not pay for it. However, spurred on by the prospect of famine, the Government authorised the collection of returns in Ireland. In Scotland statistics were collected in each of the years 1854–1857, largely through the enterprise of the Highland and Agricultural Society. Detailed figures were obtained from virtually all the Scottish farmers most of whom all too readily saw the advantage of having such information. In England small localised experiments based on the Scottish model were carried out but with much less co-operation from the farming interests. In 1855 the government announced that the results did not justify continuing with the experiment, and it was to be 11 years before problems of finding suitable machinery and motivating the farmers were overcome.

The Census

The 1801 census gives the numbers of 'persons chiefly employed in Agriculture' though there are grounds for believing that little reliance can be placed on the accuracy of the figure given for any one parish.

The 1831 census gives the number of farmers in each parish, from which the average farm size can be worked out using the acreage totals for each parish, though this method makes the (unwarranted) assumption that farm boundaries did not cross parish boundaries.

The enumerators' Schedules for the 1851 census include the names of farmers, showing how many acres they farmed and how many labourers they employed and in what capacity. The 1841 Schedules give only the names of farmers and of their farms.

4.6.2 The Agricultural Returns of 1866

The decision in 1866 to collect information for crops only had been made in principle the year before but in 1865 there was a severe outbreak of cattle plague and this disaster convinced many farmers of the desirability of having details of livestock numbers and these were in fact collected for livestock on 5 March 1866 and published on 7 May 1866. Acreage and crop figures were collected on 25 June from everyone occupying more than five acres of land. The opposition of the farmers to giving details of their crops was far greater than to giving details of their livestock. Where information was refused the enumerators were asked to estimate it, and the degree of estimation is a good guide to the farmers' lack of co-operation. In Hertfordshire 45 per cent of the crop acreage had to be estimated. Variation in response from parish to parish was considerable. For example Coppock (1956) calculated the amount of estimation in three adjacent parishes as Essendon (0), Little Berkhamstead (45%) and Bayford (74%). The opposition continued in the first few years and it was generally greatest not among the farmers but among the landlords. Inaccuracies arose at first because the enumerators were inexperienced in conducting the inquiry and because farmers interpreted questions differently. Enumerators acquired skill but there were many problems over definition of terms. It was hard to define the crops and livestock in such a way that they would mean the same thing to farmers in all parts of Britain. When new information was asked for or when existing categories were subdivided fresh problems arose. A major source of difficulty was the distinction between permanent grass and rough grazing, most difficult in Highland Britain and in the rolling chalk country of Southern Britain where there were large areas of unfenced downland. Another problem was with rotation grassland: how long did it have to lie fallow before it was considered to have reverted to permanent grassland, and how long could neglected permanent grassland escape reclassification to rough grazing? This was a pertinent question in the all-too-frequent event of a depression in farming at various times since 1866. Orchards were often returned twice, as orchards and as permanent pasture. Market gardens were a problem. First recorded in 1872, the growing practice of raising vegetables on arable farms made the old traditional meaning of market garden obsolete and the category was eventually abandoned. Woodland was usually estimated very roughly until the very large scale plans of the Ordnance Survey reached most of the country and provided, for the first time and not only for woodlands, an accurate measurement of area.

Another source of complication is the changes in farm and parish boundaries over time which could result in a holding lying in two or more parishes; as an occupier was required to make only a single return for each holding, it follows that some land would have been returned in the other parish. Since holdings tended to become larger in the course of time the degree of discordance gradually became worse.

Of the 14 926 Civil parishes recorded in the 1881 census, 3258 had their boundaries changed before 1891. Since the Government financed the project in order to get a national view of the production and the changes in production, variations from parish to parish were of no account at all to the compilers of the returns. The parish figures were the means, not the end. It is also unusual for the agricultural parish to remain unchanged for a long period and this is another source of difficulty. The total acreages given in the parish summaries often change dramatically from year to year. These may be related to changes in the boundaries, but they may also be related to significant changes in the pattern of land use, to the transfer of land from agricultural to residential use, to changes in the size of holdings and to the transfer of holdings from one parish to another. Where such sudden changes have been observed the only safe working rule is to avoid jumping to a conclusion about the reason.

Comparability from year to year is also affected by the changes made in the minimum size of holding for which returns were expected. In 1866 it was five acres. Since this size excluded many owners of livestock the lower limit was abolished altogether in 1867 and returns were required from all occupiers of land. The increase in the total amount of arable land by one million acres from 1866 to 1867 was no doubt largely explained by this abolition of the minimum size of holding. From 1869 a lower limit of ¼ acre was adopted until 1892 when the present minimum of one acre came into force.

Coppock (1956) concludes his extended examination of the June Returns with these words:

> Ratios of different crops and densities of stocking, particularly where these are based on mean values for a number of years, provide the only safe basis for comparison, particularly when counties and parishes are being studied. Comparisons from year to year can only be made with caution in view of the changes of definition and of the different interpretations which have occurred, but......most crop and livestock figures are sufficiently comparable over long periods to be of great value.

4.6.3 Access to the Agricultural Returns

The Agricultural Returns: parish summaries' (*Figure 4.7*), as they are officially known, are to be found in three different places depending on how old they are.

Summaries for the past five years only can be consulted at the Divisional Offices (normally one in each county) of the Ministry of Agriculture, Fisheries and Food.

Parish summaries for the past 10 years are housed in the Statistics Division II Executive, Unit B of the Ministry, Government Buildings, Epsom Road, Guildford, Surrey GU1 2LD. They can be consulted by writing for an appointment.

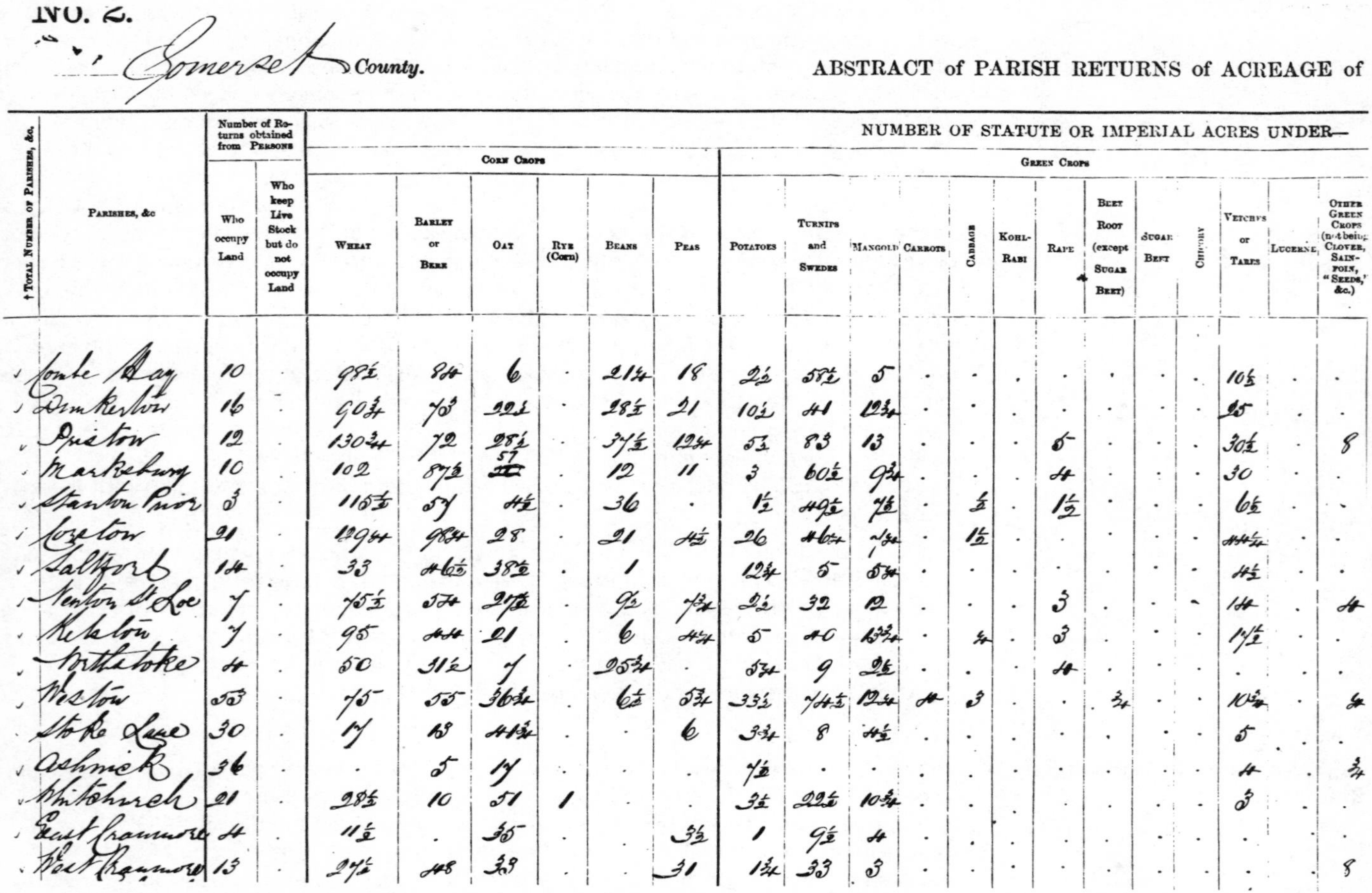

NO. 2.

Somerset County.

ABSTRACT of PARISH RETURNS of ACREAGE of

† Total Number of Parishes, &c.	Parishes, &c	Number of Returns obtained from Persons: Who occupy Land	Who keep Live Stock but do not occupy Land	Number of Statute or Imperial Acres under— Corn Crops: Wheat	Barley or Bere	Oat	Rye (Corn)	Beans	Peas	Green Crops: Potatoes	Turnips and Swedes	Mangold	Carrots	Cabbage	Kohl-Rabi	Rape	Beet Root (except Sugar Beet)	Sugar Beet	Chicory	Vetches or Tares	Lucerne	Other Green Crops (not being Clover, Sainfoin, "Seeds," &c.)
	Combe Hay	10	.	98½	84	6	.	21¼	18	2½	58½	5	.	.	.	.	.	.	.	10½	.	.
	Dunkerton	16	.	90¾	73	22½	.	28½	21	10½	41	12¾	.	.	.	.	.	.	.	25	.	.
	Priston	12	.	130¾	72	28½	.	37½	12¼	5½	83	13	.	.	.	5	.	.	.	30½	.	8
	Marksbury	10	.	102	87½	57	.	12	11	3	60½	9¾	.	.	.	4	.	.	.	30	.	.
	Stanton Prior	3	.	115½	54	4½	.	36	.	1½	49½	7½	.	½	.	1½	.	.	.	6½	.	.
	Corston	21	.	129¼	98¼	28	.	21	4½	26	46¼	7¾	.	1½	.	.	.	.	.	44¼	.	.
	Saltford	14	.	33	46½	38½	.	1		12¾	5	5¾	.	.	.	.	.	.	.	4½	.	.
	Newton St Loe	7	.	75½	54	27½	.	9½	7¾	2½	32	12	.	.	.	3	.	.	-	14	.	4
	Kelston	7	.	95	44	21	.	6	4¼	5	40	13¾	.	¼	.	3	.	.	.	17½	.	.
	North Stoke	4	.	50	31½	7	.	25¾	.	5¼	9	2½	.	.	.	4	.	.	.	.	.	.
	Weston	53	.	75	55	36¾	.	6½	5¾	33½	74½	12¼	4	3	.	.	¾	.	.	10¾	.	2
	Stoke Lane	30	.	17	13	41¾	.	.	6	3¾	8	4½	.	.	.	.	.	.	.	5	.	.
	Ashwick	36	.	.	5	17	.	.	.	7½	.	.	.	.	.	.	.	.	.	4	.	¾
	Whitchurch	21	.	28½	10	51	1	.	.	3½	22½	10¾	.	.	.	.	.	.	.	3	.	.
	East Cranmore	4	.	11½	.	35	.	.	3½	1	9½	4	.	.	.	.	.	.	.	.	.	.
	West Cranmore	13	.	27½	48	39	.	.	31	1¾	33	3	.	.	.	.	.	.	.	.	.	8

Figure 4.7 An example of the Abstract of Parish Returns. The format changed from time to time

Progressive No. 2

CROPS and of LIVE STOCK—1881

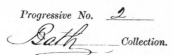

Bath Collection.

Flax	Hops	Bare Fallow or Ploughed Land from which a Crop will not be taken this Year	Clover, Sainfoin, "Seeds," Rye and other Grasses under rotation for one or more Years	Permanent Grass, as Meadow, Down, or Pasture not broken up in rotation (inclusive of Grass Land used for Fruit Trees of any kind, but exclusive of Heath or Mountain Land)	Orchards: Acreage of Land planted with Apple, Pear, Plum, Cherry or Nut Trees (Grass or other Crops growing between the Trees being also entered in other columns)	Market Gardens: Total Acreage of Land used by Market Gardeners for growth of Vegetables and other Garden Produce (to include Acreage entered under separate Crops.)	Nursery Gardens: Acreage of Land used by Nurserymen for growing Trees, Shrubs, &c.	Live Stock on 4th June, 1881 — No. of Horses (including Ponies): Used solely for purposes of Agriculture and by Market Gardeners	Unbroken Horses of any Age (including Foals)	Mares kept solely for the purpose of breeding	Number of Cows and Heifers of all ages in Milk or in Calf	Number of Cattle other than those in Milk or Calf: Two Years of Age and above	Under Two Years of Age (including calves)	Number of Sheep of all kinds, One Year Old and above	Number of Lambs Under One Year Old	Number of Pigs of all kinds and of all ages	Acreage of Woods, Woodlands, Coppices, or Plantations in 1880
.	.	50	136½	501¾	6¾	.	.	22	4	.	40	15	50	377	298	91	6
-	-	14	88	517¾	15	.	.	35	8	.	68	22	67	353	294	151	4
.	.	16	92¾	1137	25¾	.	.	46	15	1	109	70	152	780	594	160	.
.	[illegible]	.	72	1318	17¾	.	-	25	27	2	185	63	75	652	375	140	4
.	.	.	38½	467	12	.	-	19	10	.	28	51	69	439	218	25	.
.	.	30	130½	643½	19½	10½	.	40	23	3	62	43	63	487	258	69	
.	.	9	60	401¾	6¾	.	-	17	5	—	69	31	46	22	1	130	.
.	.	7	105	1047	9	.	-	26	10	1	100	72	76	459	214	106	3
.	-	14	84	629½	12	.	-	25	4	.	48	78	80	376	184	75	4
-	.	17½	32½	281½	2½	-	-	18	14	1	50	31	60	261	99	68	.
.	-	.	87	1780¼	26¼	25	2½	44	87	57	284	84	81	810	355	260	.
.	.	.	1¾	2051½	.	.	.	30	21	2	434	52	130	1	.	477	194
.	.	4	3	1466	¾	.	.	18	8	5	340	8	100	2	.	343	108½
.	.	3	55	1275½	.	.	.	30	15	1	267	14	117	380	183	396	27
.	.	8¼	59¾	603	3¼	.	.	16	2	.	130	1	42	.	.	131	264¾
.	.	.	137	1369	1½	.	.	28	13	3	298	14	191	24	4	198	39½

The parish summaries more than 10 years old are housed in the Public Record Office and they are coded under MAF 68. To examine the summaries it is necessary to apply in writing to: The Search Department, Public Record Office, Ruskin Avenue, Kew, Richmond, Surrey, asking for a reader's ticket and stating what one wants and why. For summaries after 1918 one is required to undertake that one will not extract figures for any parish with less than three farms, to preserve confidentiality.

If statistics are needed for only one or two parishes and for only a few years the Public Record Office does operate a photocopying service. If this is needed, write specifying clearly what is wanted, and a form, which is both an order form and an estimate of the cost of the copies, will be sent in return.

4.6.4 Probate inventories and other sources

A probate inventory is a list drawn up by neighbours of the possessions of a deceased person who has left a will. It lists furniture and household possessions, room by room, animals and implements in the yard, animals and crops in the fields and the quantity of grain, hemp, flax, wool, wood and hay in store. The animals are usually counted and classified with great care. Sometimes but not usually the exact acreages of the crops growing in the fields are recorded. Probate inventories begin in about 1530 and cease about 1830. The lists were made for wills proved in local probate courts and housed in local probate registries or in local record offices.

A major problem is to what extent the range of inventories represents all classes of the community. Most people died without leaving wills, but the inventories that have been studied in detail do show a broad social spectrum, whether accurate or not we do not know. Nor do we know how numerous and how poor were the very poor who were left out. Did the habit of leaving wills gradually lose its hold on the poorer classes or were there fewer poor in the eighteenth century than in the seventeenth century? These are questions that remain to be answered, but they will only be answered when more studies have been made. For our purpose the inventories are a great help in filling in the fine detail in the village community, in identifying property owners, in locating parcels of land and in gaining some insight into the social composition of the village at different times.

Another source worth exploring, usually linked with deaths but not necessarily so, is the notification of farm sales in local newspapers and by auctioneers. It was not uncommon for the executors to sell off the farm, or the stock if the farm was leased, on the death of the farmer. Notification of the sale together with details of the machinery, stock, stores and

UPPER STANTON, STANTON DREW,

SOMERSET.

WILLIAM KING

Has received instructions from the Executors of the late Mr Thomas Membery, to SELL BY AUCTION, on the Premises, on

TUESDAY, Feb. 9th, 1892,

PART OF THE

HOUSEHOLD

FURNITURE

OUT-DOOR EFFECTS,

FOWLS, &c.,

Comprising Millpuff Bed, Dressing Table, Swing Glass, Easy Chair, Sofa, Kitchen and other Tables, Stools, Sundry Ware, 3 Cider Casks, 2 Donkey Carts, Set Donkey Harness; 7 Planks, 11 by 3 inches, 13 feet long; Scaffold Poles, Ladders; Cart Shed, 30 by 9 feet; Corn Bin, Carpenter's Bench, Laths and Quartering, lot of Mason's Tools, lot Freestone, Turning Lathe, Garden Tools, Iron Boiler, pair of Grates, Rain Water Casks, Salting Stone, Grindstone, Hand Saws, Cross-cut Saws, Tunpail, Yoke, Iron Bar, Iron Pot, a few Potatoes, and other Effects.

A FEW COUPLES OF FOWLS.

SALE AT TWO O'CLOCK.

Chew Stoke, February 1st, 1892.

T. D. Taylor, Sons, and Hawkins, "Times and Mirror" Office, Bristol.

Figure 4.8 A country auctioneer's poster of 1892

equipment was often put in the local paper, or appeared as posters issued by the local auctioneer responsible for the sale. The details, which are often in effect a complete list of everything on the farm, give an excellent idea of the nature of the farm economy, and of the standard of living of the farmer and his family. Local craftsmens' possessions were also sold by auction if for any reason the beneficiaries had no use for them, or had a greater need for the money, and again the specific details, such as those shown in the poster in *Figure 4.8* have an appealing poignancy that jumps across the years.

REFERENCES

Adams, I.H. (1976). *Agrarian Landscape Terms, a Glossary for Historical Geography.* IBG Special publication No. 9.

Chambers, J.D. and Mingay, G.E. (1966). *The Agricultural Revolution, 1750–1880.* London; Batsford

Coppock, J.T. (1956). 'The statistical assessment of British agriculture.' *Agric. Hist. Rev.* **4**, 4–21 (Pt 1), 66–79 (Pt 2)

Coppock, J.T. (1965). 'The cartographic representation of British agricultural statistics.' *Geography* **50**, 101–114

Evans, E.J. (1976). *The Contentious Tithe, 1750–1850.* London; Routlege & Kegan Paul

Gray, H.L. (1915). *English Field Systems.* Cambridge, Mass; Harvard University Press

Hammond, J.L. and Hammond, B. (1966). *The Village Labourer, 1760–1832.* London; Longman

Hilton, R.H. (1955) 'The content and sources of English agrarian history before 1500.' *Agric. Hist. Rev.* **3**, 3–19 (Pt 1)

Hilton, R.H. (1976). *Peasants, Knights and Heretics.* Cambridge University Press.

Kain, R. and Prince, H. (1977). *The Tithe Surveys of England and Wales.* Folkestone, Kent; Dawson

Orwin, C.S. and Orwin, C.S. (1954). *The Open Fields,* 2nd edn. Oxford University Press

Perry, P. (1970). 'A source for agricultural history, newspaper adverts.' *The Local Historian* **9**

Prince, H.C. (1959). 'The tithe surveys of the mid-nineteenth century.' *Agric. Hist. Rev. 7,* 14—26

Tate, W.E. (1967). *The English Village Community and the Enclosure Movements.* London; Victor Gollancz

Thirsk, J. (1955). 'The content and sources of English agrarian history after 1500.' *Agric. Hist. Rev. 3,* 66—79 (Pt 2)

Young, A. (1809). *General View of the Agriculture of Oxfordshire*, pp. 35—36. Quoted in Chambers, J.D. and Mingay, G.E. (Eds) (1966). *The Agricultural Revolution 1750—1880* p.48. London; Longman

Local place and field names

English Place Name Society. A series of volumes county by county. Publication started in 1924 and still continues

Ekwall, E. (1936). *The Oxford Dictionary of English Place Names*. Oxford

Ekwall, E. (1962). *English Place Names Ending in -ing.* London; Lund Humphries

Copley, G.J. (1963). *Names and Places.* London; Phoenix House

Cameron, K. (1961). *English Place Names.* London; Batsford

Field, J. (1972) *English Field Names.* Newton Abbot, Devon; David and Charles

CHAPTER 5 POPULATION

5.I INTRODUCTION

At some stage in most historical geographical studies one needs to find out as much as possible about the population in the past. We have already seen in the sections on Domesday and in examining medieval sources such as the Lay Subsidies and the Poll Tax returns, that some evidence exists that allows us to make a guess about absolute numbers, and to make deductions, carrying rather more weight, about relative population densities between different places or areas. By the middle of the sixteenth century more sources are available. Some of them like parish registers are very detailed and very promising as data sources; others like the Compton Census of 1676 are partial counts, made at a single point in time, but leaving out whole sections of the community. By 1801 the first official census was taken, but it is not until half a century later that much more detailed census information was collected, so allowing us for the first time with some confidence to build up a fairly detailed and accurate picture of the population at a single point in time.

Nevertheless for all its defects this early material on population is of great value if one recognises and allows for its limitations, and in the following sections we shall look critically at some of the principal sources.

5.2 PARISH REGISTERS

5.2.1. Introduction

In September 1538 Cromwell ordered every parson, vicar and curate to enter in a book every wedding, christening and burial in his parish, together with the name of the parties concerned. Entries were to be made each Sunday in the presence of one of the churchwardens. At first the entries were made on paper but 60 years later it was ordered that these early entries be transcribed in books of parchment, so that the 800 or so of the 12 000 parishes whose records go back to this early period rarely contain the original entries.

In the past 20 years the enormous value of these parish registers as sources of incomparably detailed information has come to be recognised. Above all they are valuable because they cover much of the crucial period from 1538 until 1837 when modern registration of vital events such as births, marriages and deaths began.

The parish was the basic unit of local government and the one most intimately connected with the social and economic development of the country. When one looks at the original handwritten entries recording vital events in the history of the village community the first impression is of very fine detail and considerable accuracy. Detail there certainly is and there is little evidence that what is entered in the registers is actually wrong. The difficulties arise over what may have been left out.

One major source of omissions are Dissenters, people who refused Anglican communion on conscientious grounds and whose baptisms and burials often took place away from the church and its Offices. Also omitted were the adherents of the Established church who were unable or unwilling to pay the fees connected with ecclesiastical registration. As dissent grew and industrial expansion gave rise to new urban development it is fairly certain that the parish registers became even more inaccurate.

Hardwicke's Marriage Act of 1753 made it virtually impossible to contract a valid marriage unless it were in an Anglican church. On the other hand common law marriages could be and were contracted without benefit of clergy. Children dying before baptism may not be recorded, and criminals, suicides and the unbaptised were generally not buried in consecrated ground and may not be recorded in the registers. Strangers should be omitted from consideration if they can be identified as such. This is not usually very difficult because the names of travellers and strangers are often entered with a note, e.g. in Newton St Loe, the death was noted of 'a certain wandering and speechless woman' and another time of an Exeter man, who 'died on the road'.

There are two features of the burial registers that need explanation. An Act of 1666 and its successor of 1678 decreed that apart from plague victims, all bodies for burial must be wrapped in a woollen shroud and buried in a coffin, lined if at all with a woollen cloth. This was a crudely protectionist device to give a boost to English woollen production in the seventeenth century. The Act was repealed in 1814 and widely flouted for some time before that. However, failure to observe the law brought a fine and the relations of the deceased were required to certify to the fact that a woollen shroud had been used. It is not uncommon to find a capital letter M against the name of some of those buried. This signifies that 'Mortuary' had been paid. In feudal times the lord of the manor had the right to choose the best beast of the deceased person, and the parish priest to choose the second best. For the priest this was in theory a way for him to make up for any tithes the deceased may have forgotten to pay during his lifetime. In many parishes the custom lasted well into the past century.

Throughout the sixteenth century and a great part of the seventeenth 'close seasons' for marriages were observed and this may well show up in the marriage totals. The close seasons were: Advent to St Hilary's Day (13 January), Septuagesima to Low Sunday, Rogation Sunday to Trinity Sunday. The custom lapsed under Cromwell and attempts to revive it under the Restoration were not very successful but it may be a factor in explaining seasonal variations in very early sets of registers.

The amount of information given in the registers varies depending on the conscientiousness of the clergymen responsible for the entries. Some contain only the bare essentials, others are full of detail about the weather, especially extreme events, political matters and personalities. In burial entries it is not uncommon for the incumbent or his clerk to record his opinion of the deceased. At Beeston, Notts. the burial entry of Ann Parrot on 18 June 1736 has the note, 'a dissembling canting hollow presbiterian'. Rolleston, Notts. is rich in entries where the facts are given in English and the comments in Latin.

> Margerie Deconsonne the wife of Bartholomew Deconsonne...fiftie yeares of age a tall slender womā, providently thrifty, perhaps I should rather say stingy (*frugi admodū providene magis fuerit an parcior nescio*) shee leaving this life on Monday was buried on Tuesday the 30 of Aprill (1588).

> Willm̄ Forrest about 60 years of age a cūninge fellow I will not say crafty, of little faith, or hope of eternal life, if it be permissible to regard words as an index to the mind (*nō dicam versutū fide pusillū, spe vitae aeternae, si ex verbis, mentis indicibus, cōiecturā facere liceret nimium*), but in handie woorke as ditchinge, mowinge, sheip-clippinge & such like skilful: was buried December—xxviijth Tuesday.

> Aug. 12. 1687. John Wise, Bachelor, a frequenter of taverns rather than of the Church and Sacrament, attacked by a raging fever, vomiting dreadful curses and blasphemies, died and was buried. In the hour of death Good Lord deliver us. (*Johānes Wise Caelebs, cauponularum magis quam Ecclesiae et Sacramenti frequentator, febre dementi accensus, diris execrationibus et blasphemijs evomens, mortem obijt: et sepultus fuit. In hora mortis libera nos Dñe.*) (Tate, 1946, p.66)

The widespread disturbances caused by the Civil War, and the ejection of many of the clergy from their churches wrought havoc with the keeping of the registers and in some parishes records cease altogether for a while. The main gaps tend to be in the period 1642—1653 and the following extract from Kibworth, Leicestershire gives a reason for the deficiencies.

AD 1641. Know all men that the reason why little or nothing is registered from the year 1641 until the year 1649 was the civil wars between King Charles and his Parliament, which put all into a confusion till then, and neither minister nor people could quietly stay at home for one party or the other.

The responsibility for preserving the registers lay for most of the time on the resident clergyman and his churchwardens. Some took their duties more seriously than others. Registers have been found propping open church doors, half eaten by rats, and being used to prevent rain and draughts from coming though broken windows.

Growing concern at the possibility of further loss through neglect of these priceless records accounted for the Parochial Register and Records Measure of 1929 under which the bishop of a diocese was empowered to give any directions necessary to ensure the proper preservation of all records in the possession of the parish authorities. Dioceses were encouraged to set up record offices in which vulnerable records could be safely housed. However, the proportion of parish records held in diocesan record offices varies considerably, and when searching for the registers it is usually best to start in the parish itself. When the registers are in the church they are legally in the custody of the incumbent and he may if he wishes charge a fee for each year consulted. Generally however clergy are most generous in permitting access to the registers to bona fide students but they are not bound to permit access, and where it has been allowed, it is not unreasonable to make a contribution to church funds or to offer to make available to the church the results of the research.

There should be few difficulties in interpreting the entries in the registers, even though on occasion the handwriting may prove rather hard to follow. In many cases registers have been transcribed and even printed, and it is worth checking on this before starting a detailed analysis of the original registers.

Where substantial gaps appear in the registers one can sometimes make use of the bishops' transcripts. After 1579 clergy were required to supply their bishop with regular transcripts of the register entries, though many often failed to do so. Today these transcripts, though lacking detail, can sometimes be used to bridge gaps in the registers. Diocesan boundaries have changed over time and it is not uncommon for certain parishes to change from one diocese to another and one needs to bear this in mind when searching for an elusive set of records. The bishops' transcripts are often held now in the county record offices, but where they are still held by the diocesan record office it should be noted that the diocesan registrars, who look after the records, are usually solicitors and are paid fees for so doing and there may well be a fee for consulting the transcripts in such cases.

5.2.2 Analysis of registers

There are two forms of analysis: aggregative analysis and nominative analysis.

Aggregative analysis involves counting up all the entries: baptisms, marriages and burials. It does not tell us as much about the population as the other method but it is simpler, quicker and more easily undertaken and we shall concentrate on this method in this section.

Nominative analysis involves bringing together material using the names of individuals as the link. Family reconstitution is the best known form of nominative analysis. This method involves recording the names of all those baptised and tracing them among the marriages 15 to 50 years later and among the burials up to 105 years later. It is thus possible to build up a figure of the actual number of people alive at any one time, the size and extent of the families, at what age people got married, had children and died. Such analysis however demands excellent records and a great deal of time and is outside the scope of this book.

Aggregative analysis

Extracting the data, choosing a parish. Where a long run of registers is available it can be a time-consuming business to extract all the data. The first consideration is how much help is available. If a group effort is possible then one can be more ambitious, but if one is on one's own then it may be that one has to choose between dealing with a short period in depth or a longer period in a more limited way. On balance if the choice is indeed in those terms then it is probably best to deal with the shorter period in depth.

In choosing a parish it is as well to bear in mind that as a general rule one needs a register with at least 100 entries a year. It is better to cover a large and populous parish for a short period than a small one for a long period. This is because where the total entries fall well below 100 a year it is almost impossible in some cases to tell whether changes in yearly totals are due to under-registration or to actual changes in the number of events being recorded. One may be able to overcome this problem of small population size by combining several small parishes, preferably contiguous, thereby getting a picture of what is happening over an area rather than in a single place. This makes sense because there was often a lot of social activity between neighbouring parishes.

It is also necessary to decide when a year begins and ends. When the registers began, the Church year ran from 25 March to 24 March. This is useful since burials, which tend to peak in the winter months, will fall in a single year. Some workers use the 'harvest year' Michaelmas to Michaelmas (29 September to 28 September), or 1 August to 31 July. It does not matter greatly which 'year' is chosen as long as one is as consistent as possible. For data spanning a long period it may be necessary to convert one kind of year into another to maintain consistency.

Finally there is the vexing problem of what to do about 'missing data', for example, registers with longer or shorter gaps in them, the identification of and correction for under-registration of all kinds, delayed baptism etc. If there is any reason to suppose that there may be such a problem with a particular set of registers one should consult Drake (1974).

Recording the data The forms reproduced in *Figure 5.1* were devised by the Cambridge Group for the History of Population and Social Structure, led by E.A. Wrigley and Peter Laslett. For a preliminary analysis a less complicated form of recording may well be appropriate. It is important at all times to be consistent, clear and accurate in making a record of the register data. Do not change the method of recording once you have started.

Displaying the data Given a large enough data set it will soon become apparent that there are both seasonal and longer term periodic variations in baptisms, marriages and burials. An initial plot of the raw figures should be made first because patterns are usually more clearly suggested by a graph than by a column of figures. At some stage it may well be important to smooth out short term or random fluctuations. A moving mean, spanning an odd number of years will do this successfully. It is simple if at times tedious to calculate. If an 11 year moving mean is selected as suitably long to iron out the short term fluctuations, the first step is to add up the totals for the first 11 years and to enter the result under the middle year, in this case under the sixth year. Then the 11 years from year 2 till year 12 are totalled, divided by 11 and the result entered under the 7th year. The process is repeated until the last group of 11 years in the sequence has been averaged and then it has to stop. So the moving mean always traces a path on the graph that falls short of the beginning and of the end of the run of figures to which it refers.

Figure 5.2 shows the results of plotting an 11 year moving mean of the baptisms and burials for the small village of Newton St Loe. The total number of events recorded each year in this parish makes it far too small for further analysis, but none the less the directions followed by the moving means show a consistent and not unreasonable trend in the period covered by the registers.

P.E.F. I

PARISH:

County:

YEARS:

YEAR	MONTH OF CONCEPTION												Baptisms by Civil Year (totals)			Concep- tions by Harvest Year (totals)	Bastards	Comments
	Apr.	May	June	July	Aug.	Sept.	Oct.	Nov.	Dec.	Jan.	Feb.	Mar.		Jan.-Apr.	May-Dec.			
	MONTH OF BAPTISM																	
	Jan.	Feb.	Mar.	Apr.	May	June	July	Aug.	Sept.	Oct.	Nov.	Dec.						

MARRIAGES

P.E.F. III

PARISH:

County:

YEARS:

YEAR	Jan.	Feb.	Mar.	Apr.	May	June	July	Aug.	Sept.	Oct.	Nov.	Dec.	Civil Year (totals)	Jan.-July	Aug.-Dec.	Harvest Year (totals)	Comments

BURIALS

P.E.F. II

PARISH:

County:

YEARS:

YEAR	Jan.	Feb.	Mar.	Apr.	May	June	July	Aug.	Sept.	Oct.	Nov.	Dec.	Civil Year (totals)	Jan.-July	Aug.-Dec.	Harvest Year (totals)	Wanderers	Comments

Figure 5.1 Forms used for recording data from parish registers

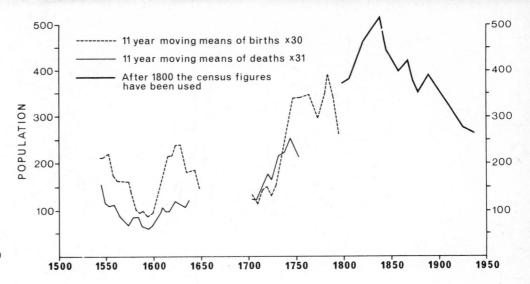

Figure 5.2 Baptisms and burials for Newton St Loe, plotted on the basis of an 11 year moving mean, 1546—1650 and 1700—1801

Figure 5.3 shows an unsmoothed set compiled by Miss Pamela Sammons of the Department of Geography in the University of Bristol, based on the more extensive registers of the church of St Mary, Harrow-on-the-Hill, between 1560 and 1640.

It is most important always to realise that the only hard information we have is the actual figures of baptisms, marriages and burials. At no stage can we be certain we know the actual number of people living in a parish at a particular time. There are various partial estimates made at different times as we have seen but these are imperfect and subject to considerable errors. It used to be argued that during the seventeenth and early eighteenth century the death rate was fairly constant at 32 per 1000 or roughly 1 in 31. If this were indeed the case then multiplying the number of burials by 31 should give a rough guide to the total population. Multiplying the births by 30 was thought to be equally appropriate. The graph for Newton St Loe has been compiled using these constants and, in addition, census totals since 1800 have been plotted. It could be argued that on the whole the match between the register data and the census data is fairly satisfactory. But is it? An interpretation of the

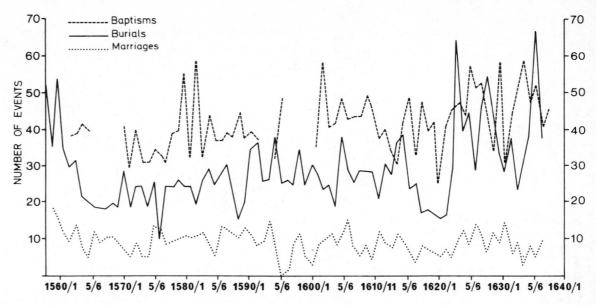

Figure 5.3 St Mary, Harrow-on-the-Hill. Fluctuations in burials, baptisms and marriages (by harvest years) (After Sammons, 1976)

early part of the graph based on baptisms shows a double peak with peaks in 1550 and 1630 and troughs in 1600 and 1650. From 1710 there appears to be a steady increase in population. But in fact we have assumed constant birth and death rates and applied these rates to actual births and deaths. When we get a fluctuation in the consequent graph we assume an increase or decrease in the total population. In fact it might very well be that the rates were not constant but fluctuated and that the population in reality grew, if it grew at all, slowly but steadily all the time. This may not be likely but it is no more unlikely than that the vital rates were constant. So in all cases it is best not to attempt to guess at total population, not to work out birth and death rates from a single population figure, or from a partial estimate that may happen to have survived.

Another difficulty is caused by migration. If a parish population were totally self-contained then it would be easier to eliminate many of the problems, but in fact people moved in and out of parishes all of the time and there are few reliable means of estimating

TABLE 5.1 BURIALS, ST MARY, HARROW-ON-THE-HILL 1570—1589 (After Sammons, 1976)

Year	X	$\bar{X}$	$(X - \bar{X})$	$(X - \bar{X})^2$
1570	43	38	5	25
1571	29	"	−9	81
1572	40	"	2	4
1573	31	"	−7	49
1574	31	"	−7	49
1575	35	"	−3	9
1576	32	"	−6	36
1577	30	"	−8	64
1578	39	"	1	1
1579	40	"	2	4
1580	55	"	17	289
1581	32	"	−6	36
1582	58	"	20	400
1583	32	"	−6	36
1584	46	"	8	64
1585	35	"	−3	9
1586	35	"	−3	9
1587	39	"	1	1
1588	37	"	−1	1
1589	46	"	8	64
	765			1231

the numbers involved. If one has been able to study a number of neighbouring parishes some problems are less difficult. If for example many of them show the same pattern of burials and baptisms it becomes less likely that in every case the pattern was caused by in or out migration.

Simple computations from register material A series of baptisms, marriages and burials can be statistically analysed using simple methods.

Figure 5.3 shows graphically the variations in baptisms, burials and marriages in the parish of St Mary, Harrow-on-the-Hill from 1558 until 1637. It is clear that there were considerable variations from year to year and we need some way to indicate how much variation there was from year to year in each of the three measures.

The first statistic needed, and the easiest to compute, is the mean. *Table 5.1* shows the burial totals for each year in the parish from 1570 until 1589. The mean is found by dividing the sum of each yearly total (ΣX) by the number of years (N = 20). This involves dividing 765 by 20 = 38.25. So the mean ($\bar{X}$) = 38.25.

The mean gives no indication however of the variation from year to year. To express this we need to find the standard deviation. This is the average amount of deviation from the mean and is calculated by taking the square root of the average of the squares of the deviation from the mean. The formula used is:

$$\text{s.d.} = \sqrt{\frac{(X - \bar{X})^2}{N}}$$

The last column of *Table 5.1* shows $(X - \bar{X})^2$, the sum of which is 1231. 1231 ÷ 20 = 61.55 and the square root of 61.55 is 7.84. So the standard deviation in this example is 7.84.

In practice we can use the standard deviation, which defines the limits of a band above and below the mean, to define extremes. Values that exceed or fall short of the upper and lower limits of the band can be considered more or less abnormal. This is the method used later to define 'crisis mortality'.

The other statistic that is useful in this context is the coefficient of variation, which shows how far a series is represented by the mean. A low coefficient of variation shows that most points in the series lie fairly close to the mean. A high coefficient of variation shows that many points lie at some distance on either side of the mean value.

The coefficient of variation (V) equals the standard deviation divided by the mean, or:

$$V = \frac{\text{s.d.}}{\overline{X}}$$

In the example in *Table 5.1* $V = \dfrac{7.84}{38.25} = 0.204$.

TABLE 5.2 SELECTED STATISTICS FROM THE REGISTERS OF ST MARY, HARROW-ON-THE-HILL (After Sammons, 1976)

	$\overline{X}$	s. d.	V
Burials	28.73	10.44	0.363
Marriages	8.98	3.76	0.417
Baptisms	41.28	7.97	0.193
Spring burials	7.18	4.12	0.573
Summer burials	6.19	3.64	0.588
Harvest conceptions	10.54	3.35	0.317

One advantage of using the coefficient of variation is that it gives an index of relative variability. It does this by eliminating the direct influence of the value of the mean. So one can use it to compare the variability of a series of baptisms, marriages and burials even though there may be great variation in the average value of each of the three sets of figures.

Using 'harvest years', i.e. from the beginning of August in any one year to the end of July in the following year, and spanning the period 1558 till 1639, the registers of St Mary, Harrow-on-the-Hill can be made to yield the information in *Table 5.2*. Reference to this table and to *Figure 5.3* throws some light on the demographic situation in the parish.

The arithmetic mean for baptisms (41.28) is substantially greater than that for burials (28.73). Burials in fact exceeded baptisms on only three separate occasions. The number of deaths appears to fluctuate more than births from year to year, with V values respectively of 0.363 and 0.193.

The indications are very strong that the population was probably increasing over the period, but as we saw earlier we cannot be sure of this since any potential increase might have been offset by emigration, about which these figures tell us nothing at all.

The figures of baptisms representing conceptions that took place in the harvest period, from August to October, show a higher degree of variability (0.317) than the figures of baptisms for the year as a whole (0.193).

It is sometimes worth distinguishing burial totals for different periods to see whether there is any significant seasonal variation. In this case the pattern of spring (April to June) and summer (July to September) burials seems broadly similar and about half the burials

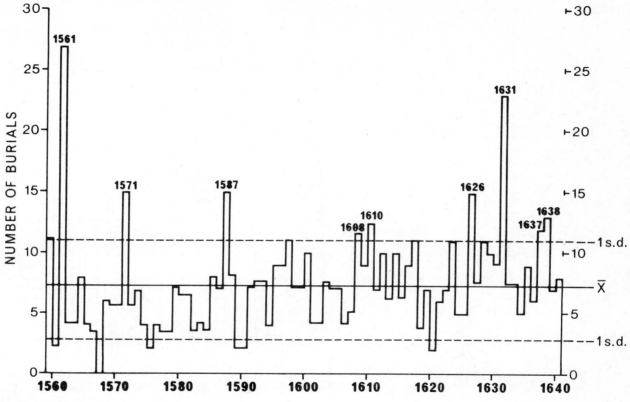

Figure 5.4 Variations in 'spring' burials, St Mary, Harrow-on-the-Hill

seem on average to have taken place in this half of the year. However the higher *V* values suggest some greater variations from year to year in the seasonal totals than in the annual totals. The variations in and between the 'spring burials' and the 'summer burials' are shown in *Figures 5.4* and *5.5* respectively.

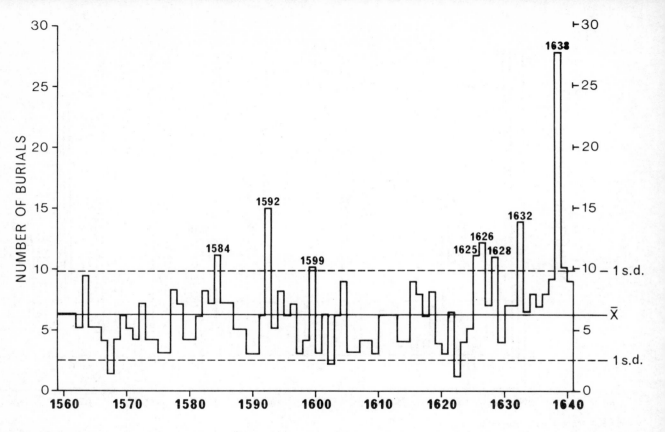

Figure 5.5 Variations in 'summer' burials, St Mary, Harrow-on-the-Hill

'Crisis mortality' This is the term given to years with unusually large numbers of deaths, say twice the average for the period. It has been argued that changes in mortality have gone a long way to determining the course of population change in the past. The crisis may be caused by a sudden reduction in the amount of food available, generally by poor harvests, or by epidemic diseases of more than usual virulence.

It is worth looking carefully at the frequency and timing of years of crisis mortality. It is now known that such years or periods were frequent in the mid-seventeenth century and had become less common by the mid-eighteenth century. Some parishes escaped them altogether; others suffered them well into the mid-eighteenth century. Well-known epidemic years were 1557–1603, 1623, 1665, 1728–1729. Often the seasonal distribution of burials in crisis years is different from the seasonal distribution in normal years. Indeed some of the big killer diseases had characteristic seasonal profiles, e.g. bubonic plague deaths often peaked sharply in late summer and fell rapidly with the onset of winter in October and November. In the 20 years after the Great Plague of London in 1665, smallpox, typhus, influenza, diphtheria, malarial ague, spotted fever and dysentry came down 'like rain through a leaky roof' as the author of Piers Plowman wrote of successive epidemics after the Black Death. Some villages remote from trade routes escaped. While some years were almost universally bad from an epidemic point of view there were very great local and regional variations, and it is always possible that local epidemics cut a swathe through a particular parish community at a time when the region or the country as a whole was relatively unaffected.

Calculation of crisis mortality is best done by computing the standard deviation of the series of burial figures. It has been suggested that where a register shows an excessive number of burials per annum and where a monthly analysis reveals that over 50% of these occur within three successive months of the 'plague season', June to October, the record is suggestive of a plague outbreak.

Marriage registers and marriage horizons

Marriage registers can be used to trace the distance over which people found their marriage partners, what is rather charmingly known as the 'marriage horizon'. The registers allow one to find the village or town where the husband lived, since it was normally the custom for the marriage to take place in the bride's parish. Given sufficient resources one can then hope to find out whether the home of the newly married couple was set up in the bride's or the groom's village or town.

One of the earliest studies of this kind was made by Peel (1941). Four villages in Northampton were chosen: Maidwell and Lamport on the main road between Northampton and Market Harborough, the others neighbouring villages of Great Everdon and Farthingstone in a more isolated position on the low divide between the Nene and Ouse valleys. Both groups are about the same distance from Northampton. For each village the registers were used to extract details of:

(a) total number of marriages celebrated in the village;
(b) place of residence of each man and each woman mentioned.

Data were recorded in five year intervals, and distance intervals were (a) within five miles; (b) 5—20 miles; and (c) over 20 miles. To allow for great fluctuations in numbers of marriages over each five year period the numbers of men and women in each distance group were computed as percentages of the total number married.

Comparison of the two groups of villages showed that in those on the main road a high proportion of men married in the seventeenth century came from the immediate locality. The

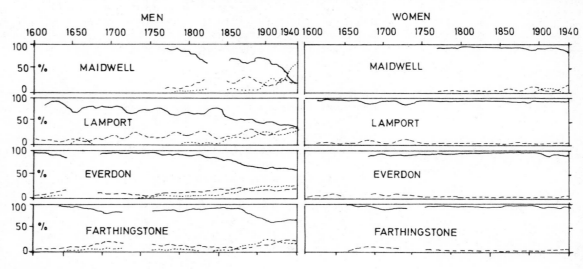

Figure 5.6 Distance effect on the selection of marriage partners. Percentage resident within: 5 mile radius (——): 5—20 mile radius (- - - -); beyond 20 mile radius (· · · ·) (After Peel, 1941)

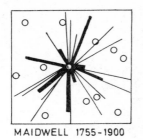 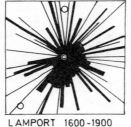 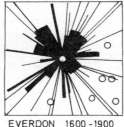 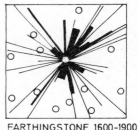

| MAIDWELL 1755-1900 | LAMPORT 1600-1900 | EVERDON 1600-1900 | FARTHINGSTONE 1600-1900 |

Figure 5.7 Direction bias in the selection of marriage partners. Thickness of line is proportional to number of marriages; O indicates village with no recorded marriage. North is at the top of each square; the side of each square represents approximately 10 miles

eighteenth century saw a slight breaking down of this rigid localism. By 1860 the percentage of local marriages had fallen steeply to 50 per cent and by 1900 strangers from over 20 miles away formed a majority of the husbands selected by the local women.

In the two more isolated villages, gaps in the registers weaken the strength of the conclusions but in general they show the same trends but in a much less pronounced form, reflecting the lesser influence of communications no doubt (*Figure 5.6*). Peel went on to analyse the marriage patterns on a spatial basis, using diagrams rather like wind roses (*Figure 5.7*). These raise many interesting questions about the directions from which marriage partners were selected. Great Everdon married especially northwards and Farthingstone shows a slight bias towards the northeast. The discrepancies may have no significance or they may be explained by long-hallowed local rivalries. At any event studies of this kind reveal spatial patterns of relationships that may for example have a definite bearing on the problem of inbreeding or the persistence of physical type. If this has to even a small degree occurred in the relatively open area of the English Midlands how much greater it may have been in the peripheral and more isolated areas in the rest of the country.

5.3 OVERSEERS OF THE POOR AND CHURCHWARDENS' ACCOUNTS

5.3.1 Introduction

The office of overseer of the poor was created in 1597—1598 and the great Poor Law of 1601 was the foundation of poor law administration for more than 200 years. An important provision of the act was that the churchwardens and up to four substantial householders in each parish were nominated each year as overseers of the poor, charged with maintaining the poor and setting them to work where possible. The funds for this were raised by taxing every inhabitant who owned land, houses, minerals or saleable timber. A further act of 1662 was the foundation of the system of settlement and removal, containing the iniquitous provision that no stranger be allowed to settle in a parish unless he rent a house or provide a security to reimburse parish officials should they be put to any expense in providing for him in the future. Temporary visitors needed a certificate from their parish guaranteeing to take them back in due course.

To counter the constant abuses of the system the overseers were required in an act of 1743—1744 to keep detailed accounts. The overseers accounts of Newton St Loe antedate the act by four years and provide large quantities of material invaluable for studying the social conditions of the period. The parochial system was wound up by the act of 1834 under which parishes were compulsorily amalgamated into Unions, each Union having a workhouse for the poor of the constituent parishes.

The period covered by the 1601 act was a period in which the problems of population, poverty and economic development were constantly proving vexing and there was much argument about the causes of poverty and even more about how it should be tackled. Some of the problems are clearly highlighted in the overseers' accounts of Newton St Loe and the following extracts are typical of those that will be found in most sets of accounts.

When an impoverished stranger came to the parish every effort was made to find out where he came from and to send him back so that the expense of supporting him should not fall on the parish. In one case in Newton there is such a complete and detailed account of the cost involved in the removal of one family that it is worth quoting in full.

Memorandum. It is agreed at a vestry held this 29th August, 1769 that Mary White, widow, and her three children who are brought to this parish from the parish of Siston, Glos. shall receive of this parish 4/- a week until it appears to what parish she properly belongs.

	£	s.	d.
Examination warrant and councellor's opinion about J. White's family	0	15	6

Sundry expenses on account of the settlement of J. White's family viz:

	s.	d.
A journey to Siston		3	0
Councellor and attorney's fees at Glos. where this parish appealed the justice's order	2	11	10
Horse hire and expense of going to Glos.		18	9

Expenses of the appeal viz:

		s.	d.
Mr. Stout attorney's bill of costs	6	16	6
Ed King for a chaise and pair of horses	1	17	6
Gave the driver		5	0
Horse hire for evidence		16	6
Pd John Turner for loss of time		5	0
Mr Balnes expense		2	9
A woman for taking care of White's child while its mother was at Glos.		1	0
Turnpikes and expenses of the journey	4	18	3
Total	18	15	2

The total expenditure on the poor in the year 1769, when the costs were incurred, was £56 5s. 2¼d. so this one case claimed one-third of the total cost of the poor in that year.

There are in the parish records several copies of the examination which paupers requiring settlement were made to undergo, and they are a very useful record as biographies of a class of person whose lives generally are so neglected.

5.3.2 Bastardy

In the period covered by the records bastardy was so common that it occasioned little attention. It was however one of the features of the settlement laws that a bastard, being unable to take the settlement of its father, was legally settled in the parish of its birth, an arrangement that was naturally not very much to the liking of the parish concerned. Here incidentally we may mention a clause inserted in the regulations in Newton regarding the appointment of a deputy waywarden in 1832. The inhabitants stipulated that

The deputy shall strictly observe and report to the vestry if any person or persons of suspicious character come into the parish and also if there be any unmarried female that may be likely to become chargeable to the parish, and see that proper steps are taken to prevent their gaining a settlement.

The following extracts from the records show that it was the concern of the officials of the parish to settle a particular case of bastardy as cheaply as possible and they therefore had to decide the more expedient course, either to force the father, if the mother could swear to him, to marry the mother before the child's birth, or merely to force the father to provide for the woman and child through some sort of bond, of which there are several in the Chest. The most usual course was the former.

	£	s.	d.
1785 Mary Cowley's examination touching her pregnancy		1	0
1811 That the present overseer do get a summons for Ann Lansdown to swear to the father of her bastard child			
1763 Expenses concerning Mary Bell with child of a bastard, going to Justice for a Warrant to take the father, and midwife and substitute in lying in. Expenses to Sessions after birth of child. To take her examination and taking out a warrant to apprehend ye father		18	0
1772 Expense in apprehending John Boswell, cost of a licence for his marriage, a ring and clerk's fees etc.	2	4	8

A recent study called *Family Life and Illicit Love in Earlier Generations* by Laslett (1977), the co-director of the Cambridge Group for the History of Population and Social Structure, looks at the changing patterns of family forms through family reconstruction and other material gained from the registers. It emerges from this study that certain English regions, such as the West country and the Northwest, appear for many centuries to have had higher bastardy rates than other regions. What the explanation for this regional variation can be we may never know, but it is interesting that it has recently been identified as a problem.

5.3.3 Apprenticeship

Apprenticeships were of two classes: those apprenticed by voluntary consent without the influence of parish officials; and the parish apprentices, bound into their trade by the parish authorities. The records kept in the parish Chest are usually but not always of the latter class. The agreement between the adults, the craftsman to whom the child was apprenticed and the parish authorities, was originally torn into three parts, the distinctive irregular edges (indents) identifying the three parts as belonging to the same document. Until 1757 this form of deed was the only one legally acceptable but afterwards any deed properly stamped and witnessed was legal.

Boys were apprenticed at 15 years of age and girls at 12. Most apprenticeships lasted for seven years but some only lasted for five.

Where they have been preserved in the parish Chest the indentures are a good source of information about the range of crafts and trades that was available and the distances that the young people travelled from their home villages. For many of them it would be the first time away from the village and it was not uncommon for marriages to be contracted at the end of the period. The mobility caused by the system no doubt helped to break down the isolation of many rural communities and reduce the element of inbreeding that was always present to some degree.

It must be admitted that by the eighteenth century it had become common to use apprenticeship as a device of getting rid of pauper children, often into some other parish. Cutlack, quoted by Tate (1946), analysed 216 apprenticeship bonds from Gnossall, Staffs. from 1691 to 1816. He was able to show that of the 178 children apprenticed within the parish, no less than 157 were apprenticed, if girls, to 'housewifery' and, if boys, to 'husbandry', i.e. to domestic service and to farm labouring respectively. From 1817 to 1835, 240 more were apprenticed but only 10 to trades in the proper sense of the term. The records of mills in manufacturing areas that imported pauper children from London give horrifying details of the abuses of pauper apprenticeship. Tate quotes Karl Marx: 'A great deal of capital was yesterday in England the capitalised blood of children.'

The indentures for Newton St Loe children in the period 1840—1866 are extant and show well not only the range of trades practised but also the distances the young people were sent. This information has been mapped in *Figure 5.8*. A systematic study of the indentures for a number of adjoining parishes over a reasonable time-span would be a most valuable and interesting contribution to an understanding of the way that the system operated.

The evidence from Newton St Loe is of a reasonably humane system, though not without

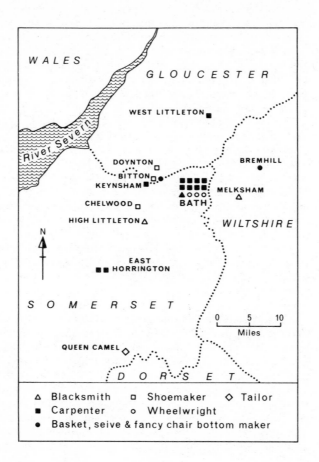

Figure 5.8 Apprenticeships arranged from Newton St Loe, 1840–1866

its problems, as this earlier extract from the Quarter Sessions Records at Wells, Somerset in January 1672 suggests:

> Order on a dispute between the parishes of Newton St Loe and Buckland that Ann Biggs be absolutely settled at Buckland unless the parishioners thereof give the churchwardens and overseers of Newton sufficient discharge for her; it being alleged that she came from Buckland to Newton as an apprentice of Joseph Biggs of Newton, Bodice Maker, who was to receive 12d a week for her maintenance from the overseers of Buckland, and that she remained there for 20 weeks and that Joseph Biggs thereafter proved insolvent and unable to take on an apprentice.

5.3.4 General and financial

Attitudes of parish officials to their responsibilities varied as widely as the characters of the officials and of the parishioners whom they were supposed to serve. In the case of Newton one is constantly struck by the broad humanity and even, on occasions, the generosity showed by the overseers.

The very detailed accounts summarise many small tragedies, for example:

> 1823. June 30th. For Betty Maggs, Calico 6/9, Blanket 4/-, Two shifts 4/6, 2 Handkerchiefs 1/9, Bed and carriage 13/-, Washing and burning filth 3/-.

Later that same year....

> Pd. a woman to look after Betty Maggs, 11 weeks at 2/6. Pd. washing and putt in coffin, liquor etc. 2/-. Bell and grave 4/-. Coffin etc. £1.1.0. ditto 4 men 4/-. Clerk for Maggs and Hunt 2/-.

The records fairly frequently mention strangers passing through and there are often accounts of payments to women in distress on the road and to soldiers with passes. On the other hand we sometimes read of rogues and vagabonds being whipped or taken to the County Gaol. There is one case in 1774 when 'a poor man whose leg was fractured by a waggon running over him' cost the parish no less than £41 14s. 8½d. in surgeons' bills and three months in the local inn. This large sum spent on a total stranger amounted to no less than a half of the total expenditure on the poor for that year. Admittedly the later entries suggest that the authorities were doing their best to get a settlement for the man in Holt in Wiltshire, but since no money seems to have come into the accounts from this source the overseers will have to be content that later generations if not their contemporaries commend them for their benevolence.

The overseers' and Churchwardens' accounts then offer a useful and, on occasion, an entertaining glimpse into the life of the ordinary villager. They can also be used to show the way in which social and economic changes were occurring at this local level.

Figure 5.9 shows the total amount raised by the poor rate from 1746 until 1834, and the annual expenditure. It also shows the numbers of people receiving weekly relief from year to year and the proportion of the total expenditure that these regular payments absorbed each year. It is obvious that the cost of supporting the poor increased steadily. Even in this more than normally prosperous parish conditions were deteriorating and it is not difficult to see why by 1834 some more radical changes in the Poor Law administration were considered necessary.

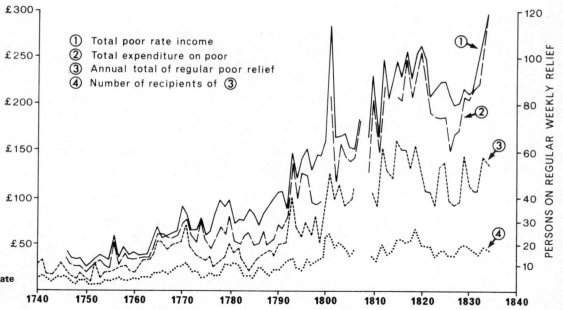

Figure 5.9 Details of the poor rate in Newton St Loe, 1746–1836

① Total poor rate income
② Total expenditure on poor
③ Annual total of regular poor relief
④ Number of recipients of ③

5.4 PARTIAL COUNTS PRE-CENSUS

5.4.1 Introduction

Parish registers for all their intimate detail and fascination as social documents cannot, as we have seen, tell us either how many people were alive in the parish at any one point in time, nor how the individuals were linked together in family groups.

The first official census was taken in 1801 but before that lists of individuals in the parish were made for specific purposes. It is not at all uncommon to discover many such lists in which a parish one is studying is mentioned but because the basis upon which such lists were compiled varied a great deal it is useful to bear in mind some general questions to try to answer about any particular data source.

First, is the list 'real' or 'ideal'? A proper official census is 'real' because it is supposed to include everyone present at the exact moment of time in which it is taken, and to exclude everyone absent, whether or not they were in their normal place of residence when the census was taken. So in a census strangers away from home are recorded in the place where they are staying on the night of the census, and not in their home villages. An 'ideal' census describes what the census takers are led to believe is a 'normal' state of affairs. It is often difficult to tell one type from another, but obviously the accuracy of the information, and what can be inferred from it, differs greatly.

Second, remembering that some types of lists do not pretend to include everyone, we must ask what class of person is being dealt with. The well-known Compton Census, an Ecclesiastical Census taken in 1676, listed all communicants aged 16 and above, and nonconformists. Before we can translate this partial count into a figure representing the entire population we need to know by what factor to increase the recorded total to allow for the children under 16 years old.

Third, if household are recorded, are all or only certain of the members counted?

Fourth, is the relationship of individuals to the head of the household given?

We have also to remember that the information recovered from a list is *static*, that is, it refers to a single point in time.

Even with all these disadvantages though it is surprising how much material can be found and put to use, provided always that one resists the temptation to read more into such material than it warrants.

Some of the material up to the fourteenth century has already been mentioned and here we shall be concerned with some sources from then until the census of 1801.

5.4.2 The Hearth Tax Returns

Between Michaelmas 1662 and Lady Day 1689 a tax of two shillings yearly was levied on each hearth in the house of every ratepayer worth 20 shillings or more a year. For the years 1662–1666 and 1669–1674 'assessments' showing what was supposed to be collected, 'returns' showing what was collected, and 'exemption certificates' for this tax are found in Exchequer and in many county record offices. In general the Lady Day 1664 assessment seems to be the most comprehensive, including as it does lists of those chargeable and those exempt. The records provide lists of householders and the number of hearths in their houses. Broadly, the more hearths in a household the larger and the wealthier it was likely to have been. The return of the township of Wirksworth in Derbyshire is summarised in **Table 5.3.**

TABLE 5.3 HEARTH TAX RETURNS FOR THE TOWNSHIP OF WIRKSWORTH, DERBYSHIRE

| | Charged | | Not charged | |
	Entries	Hearths	Entries	Hearths
1664	232	378	217	217
1670	241	440	not shown	

Charged hearths for 1670 are made up as follows:

```
  3 houses with 8 hearths
  0   -        - 7  -
  5   -        - 6  -
  5   -        - 5  -
 13   -        - 4  -
 19   -        - 3  -
 58   -        - 2  -
135   -        - 1  -
 ───
238 (discrepancy with the total first given of 3)
```

The problem is, how many people does this return represent?

Gregory King in the late seventeenth century suggested a variety of multipliers, ranging from 6.4 for an eight hearth house to 3.7 for a single hearth house. Using these on the Wirksworth figures gives a total for charged houses of 927. There is no information on uncharged hearths for 1670 but we can assume that there would not have been any great changes between 1664 and then and we can use the earlier figures. If we assess the uncharged hearths as if they were in single hearth houses and allot them 3.7 persons each we need to add a further 803, making a total of 1730.

However, Eversley (1965), working on Worcestershire returns, suggests that most of the exemptions would have been one hearth houses occupied by pauper couples or by widows or widowers, and that a multiplier of 2 for the uncharged hearths would be far more realistic. If this is applied to Wirksworth the addition is 434 and the total becomes 1361.

In general it is easier to compute a notional total if the return, like that of Wirksworth, gives the numbers of exempt or 'not charged' hearths.

Another problem is whether to base one's calculations on the number of hearths or the number of households. If hearths then King's estimates are probably as good as we can expect, though we must remember that with the hearth as the unit of taxation people had a vested interest in the assessors underestimating rather than overestimating the number. If houses are being used then King's estimate of 4.5 persons per household is reasonable, but in detail it is known that there were wide variations from this figure especially in small rural parishes and in central parishes in large towns, where more than one household commonly lived in a single house.

5.4.3 The Muster Certificates

These were drawn up between the early sixteenth and the mid-seventeenth centuries. Those compiled under Henry VIII are generally regarded as more accurate than the Elizabethan ones. The method of raising troops for the defence of the nation during the Tudor period was to hold musters in the counties. Every free able-bodied man was expected to keep ready for use the arms appropriate to his rank and wealth and to be prepared to serve when summoned.

In practice people rarely attended a muster except in an emergency because for many the journey would have been slow, inconvenient and costly and the muster would have strained the goodwill of the local authorities in the place where it was held. The provision of armour (the furnishing of harness as it was called) was the responsibility of the parish township or city and the muster commissioners had regularly to inspect the armour and to see that everyone in their area who should do so contributed to its cost. Muster Certificates were drawn up by the commissioners, listing those who had the various types of armour, and the assessment on land and goods of those required to provide cash support. The form of the record as it has come down to us varies a little from county to county but the entry shown below for Newton St Loe in Somerset is fairly typical. The date of the Somerset muster is 1569.

Richard Gebbons	pekeman
John at Wood	archer
John Gibson	light horseman
Edward Gilson	gonner
Richard Smythe	billman
John Bartlett	gonner

Armour

One tithling corslet furnished, a coat of mail furnished.
Edward Newell Esq. ij corslets furnished, ij coats of
plate furnished, ij harquebuts and two geldings for
light horsemen furnished.

Some of the terms used in the extract as well as others may need explaining. The bill was a dual-purpose weapon, with a point at the end and a blade at the side. Archers still used the traditional longbow. Harquebut is another way of spelling arquebus, a handgun of the period. Brigander, a term sometimes used, meant body armour for foot soldiers. Almain rivets were riveted overlapping plates to protect the shoulders and thighs and were copied from German mercenaries; hence the name Almain. ij is another way of writing ii (two).

In the Muster rolls each village is listed in some detail, and from the lists one can easily add up the men counted and the amount of armour for which the community was responsible. So as with the Poll Tax returns or the Lay Subsidies of an earlier period, these records are a valuable guide to relative population and prosperity. But again some caution is needed. The 1522 certificates for Buckingham show that the valuation and assessment procedures were applied differently in different parts of the county (Chibnall, 1973). Without knowing this one might mistakenly assume that regional differences in assessment were due to differences in wealth.

Naturally it is tempting to try to work out what actual population is likely to be represented by the lists of names given in the muster returns. Unfortunately we have no certain and direct knowledge of either the age or the sex structure of the population at this time. Gregory King, working on a 1532 Muster roll and using age data he obtained in Lichfield, Staffs. in 1696, worked out that the female to male ratio was 93.4 to 100, and that 40% of the population was under the age of 16. Unfortunately we know that age structure varies according to the rates of fertility and mortality current in the previous generation and that it is especially sensitive to changes in fertility. Therefore there is no guarantee that King's figures are appropriate for either 1532 or 1569. Given this uncertainty the usual range of multipliers used is between 4 and 7. As a general rule it is far better to use the muster information purely for comparative purposes.

5.4.4 The Compton Census

The Compton Census of 1676 was not completed as consistently as the organisers hoped and the researcher would have liked. Some of the clergy who were required to make the returns sent in a list of the total inhabitants of their parish, others of the number of families, and others of just the number of communicants. Many of the clergy also understated the number of the non-conformists, possibly because they felt that too large a number might reflect on their abilities as the parish priest! A fairly general correction to apply is to multiply the returned figure by $1^2/_3$.

5.5 THE CENSUS

5.5.1 Introduction

During the eighteenth century there was lively speculation about the relations between population, economy and society. In England the burden of the poor seemed to be increasing and there was no agreement about what could or should be done about them. Emigration was seen by some to be the only answer, to others it was an indictment of the whole social system that the only way some of its members could avoid destitution was to be banished to distant corners of the empire. The mercantilists had always believed that if a nation had a large and increasing population it was bound to be at an advantage compared with smaller ones. But supposing a high proportion of the state's population was impoverished and non-productive? One major school of thought held that man was essentially good but everywhere corrupted into idleness, misery and vice by social institutions such as the government, the landowning class or the established church. Malthus, however, in 1798 argued in his famous 'Essay on the Principle of Population' that it was in the nature of things that mans' numbers would always cause him to press heavily upon the means of subsistence, and that if he were unable to exercise moral restraint (i.e. celibacy) then inevitably disease, vice and famine would keep the population within the bounds prescribed by the food supply. At the time when these spirited discussions were raging no one had any accurate information as to either how many people there were in England or whether the population was increasing or decreasing.

5.5.2 The first censuses

A systematic national count had been proposed as early as the middle of the eighteenth century. A Mr Potter introduced a Bill to that effect in Parliament but it was defeated in the House of Lords in 1753. There were strong forces at work opposing any kind of census. A number of European countries had already held censuses: Iceland in 1703, Sweden since 1748, Austria after 1754, Norway in 1769. Some of the smaller Italian and German states also had taken local rather than national counts. The mercantilists in England opposed a census on the grounds that valuable as such information might be to the philosophers and political economists, to let potential enemies know the size of one's population (especially if it turned out to be smaller than expected) would be to invite aggressive action if not actual attack. However, for many it was becoming a matter of urgency to find out not only what the actual population was but also the rate at which it was increasing or decreasing, and the first census in England was taken on 10 March 1801 (*Figure 5.10*). It revealed a total population of 9.168 million. It is known that this is an underenumeration. The enumerators were drawn from parish officials making their return and swearing to its accuracy in front of local Justices of the Peace. In fact a number of parishes failed to send in any returns at all and the enumerators' zeal would have varied a great deal from place to place. One major problem not cleared up till the 1841 census was that there was no actual nominal list of persons counted, just a single return for each parish that could not be broken down into any more detail. Some of the enumerators had great difficulty with the question about the number of families, which was widely misinterpreted.

The census was repeated in 1811; this time no parish failed to send in a return, but the category 'occupations of persons' was changed to 'occupation of families' making comparison with 1801 difficult and raising difficulties where individuals in one family had different occupations.

In the 1821 census details of age were given for the first time, but the question was a voluntary one, as can be seen by the instructions given to the enumerators.

If you are of the Opinion that in making the preceding Enquiries... the Ages of the several individuals can be obtained in a manner satisfactory to yourself, and not inconvenient to the Parties, be pleased to state (or cause to be stated) the Number of those who are under 5 years of age, of those between 5 and 10 Years of Age...distinguishing Males from Females.

HUNDRED, &c.	PARISH, TOWNSHIP, or Extra-parochial Place.		HOUSES.			PERSONS.		OCCUPATIONS.			TOTAL or PERSONS.
			Inhabited.	By how many Families occupied.	Uninhabited.	Males.	Females.	Persons chiefly employed in Agriculture.	Persons chiefly employed in Trade, Manufactures, or Handicraft.	All other Persons not comprized in the Two preceding Classes.	
Taunton Dean (continued.)	Wilton - - - -	Parish	68	74	2	139	192	148	95	88	331
	Withell Florey - -	Parish	12	15	—	43	40	24	2	57	83
			1,481	1,865	72	4,120	4,460	3,811	1,442	3,328	8,580
Tintinhull -	*Ilchester - - -	Town	136	157	2	386	431	80	128	609	817
	Kingstone - - -	Parish	28	36	1	100	97	160	33	4	197
	Montacute - - -	Parish	169	183	3	377	450	158	392	277	827
	Northover - - -	Parish	12	13	1	26	30	14	4	38	56
	Stoke under Ham -	Parish	144	172	13	362	404	90	232	444	766
	Tintinhull - - -	Parish	41	69	3	148	185	52	18	263	333
			530	630	23	1,399	1,597	554	807	1,635	2,996
Wellow - -	Camerton - - -	Parish	109	122	2	287	307	35	19	540	594
	Comb-English - -	Parish	46	56	—	112	114	156	6	64	226
	Do - Hay - -	Parish	37	43	—	139	93	29	14	189	232
	Corston - - -	Parish	50	58	—	126	142	54	15	199	268
	Dunkerton - - -	Parish	46	56	1	124	114	22	92	124	238
	Farleigh - - -	Parish	24	31	1	82	85	28	21	118	167
	Forscot - - -	Parish	13	15	—	61	39	15	4	81	100
	Hinton, Charter House - - -	Parish	135	142	12	308	311	81	123	415	619
	Newton, St. Leo -	Parish	49	69	—	186	185	27	9	335	371
	Norton, St. Phllip's -	Parish	110	117	11	268	289	115	75	367	557
	Tellesford - -	Parish	26	27	2	75	78	29	68	56	153
	Twiverton - -	Parish	105	120	4	357	407	37	310	417	764
	Wellow - - -	Parish	147	148	12	397	373	98	45	627	770
			897	1,004	47	2,522	2,537	726	801	3,532	5,059

Figure 5.10 Part of the 1801 Census for Somerset

In 1831 enumerators were asked only to specify the number of males aged over 20 years. Only in 1851 was age defined as 'age last birthday'. In 1831 the enumerators were given 'tally sheets' to help them record the information and improve accuracy.

5.5.3 System of civil registration

By 1841 the census had gained greatly in accuracy and this was due to the fact that on 1 July 1837 a system of civil registration was introduced whereby all births, marriages and deaths were recorded in registers under the supervision of the Registrar General of England and Wales. The entire country was divided into regions and these in turn were divided into registration districts and subdistricts under the supervision of superintendent registrars and registrars respectively. Every quarter the superintendent registrar sent copies of all the registered events for the preceding three months to the office of the Registrar General in London. The first Registrar General was Thomas Lister. He was quick to enlist the support and co-operation of the medical profession and in 1839 engaged Dr William Farr to analyse the vital statistics that came in from all over the country. The willingness of the doctors to co-operate by giving the cause of death which was entered in the death registers along with the sex, age and profession of the deceased was a particularly valuable departure as was noted in a contemporary document.

If therefore the cause of death be correctly inserted there will exist thenceforward public documents from whence may be derived a more accurate knowledge, not only of the comparative prevalence of various mortal diseases, as regards the whole of England and Wales, but also of localities in which they respectively prevail, and the sex, age and conditions of life, which each principally affects.

By 1841 the registrar assumed responsibility for taking the census. Perhaps the greatest step forward was the introduction in that year of a household schedule, with details of each individual in the household being separately recorded, and the enumeration as a whole relating to each individual present on the actual night of the census.

An illustration of a typical enumerator's return taken from a sample issued for the 1851 census is given in *Figure 5.11*. The best description of the procedure for compiling the schedules into books is given in the 1851 Census Report, population tables, Part I, Vol. I, Report, number of inhabitants, etc.: House of Commons 1852–3 (1631) LXXXV pp ix-xxiv. A sample page of the 1851 census is shown in *Figure 5.12*.

Parish or Township of St. James, Westminster.		Ecclesiastical District of		City or Borough of Westminster.			Town of	Village of	
No. of House holder's Schedule	Name of Street, Place, or Road, and Name or No. of House.	Name and Surname of each Person who abode in the house on the Night of the 30th March, 1851.	Relation to Head of Family.	Condition.	Age of Males.	Females.	Rank, Profession, or Occupation.	Where Born.	Whether Blind, or Deaf-and-Dumb.
4	7, Charlotte Street	Michael Mingen	Head	Mar.	20		Victualler	Ireland.	
		Mary Do.	Wife	Mar.		30		Ireland.	
		Ellen Do.	Daur.			7 m.		Middlesex ; St. James, Westm.	
		Catherine Fox	Serv.	U.		30	General Serv.	Hants ; Andover.	
		Catherine Doyle	Serv.	U.		25	Barmaid	Ireland.	
5	8, Charlotte Street	Lambert Lacken	Head	Mar.	30		Tea-dealer ; (master, employing one man)	Cumberland ; Wigton.	
		Emma Do.	Wife	Mar.		30		Cumberland ; Longtown.	
		William Do.	Son		2			Middlesex ; St. Jas. Wes r.	deaf & dumb
		Henrietta Do.	Daur.			4 m.		Do. Do.	
		George Betts	Shopman	U.	19		Tea-dealer's Shopman	Do. Shoreditch.	
		Jane Cook	Serv.	U.		22	General Serv.	Do. Mary-le-bone.	
6		James Phillips	Head	Mar.	40		Plumber	Yorkshire ; Leeds.	
		Harriet Do.	Wife	Mar.		29		Do. Do.	
		Sophia White	Serv.	U.		17	General Serv.	Middlesex ; St. Jas. Westmr.	
	Three Houses uninhabited								
7	2, Bird Lane	William Frampton	Head	Mar.	72		Coach Trimmer	Staffordshire ; Bilston.	
		Anne Do.	Wife	Mar.		74		Do. Do.	
8	3, Bird Lane	Thomas Johnson	Head	Widr.	68		Retired Grocer	Devonshire ; Honiton.	
		Emma Do.	Niece	U.		41	Corset Maker	Middlesex ; St. Pancras.	
		Jane Farmer	Apprentice	U.		18	Corset Maker (App.)	Middlesex ; Stepney.	
Total of Houses	I 4 ; U 1 ; B — .			Total of Males and Females –	7	12			

Figure 5.11 Sample page of an enumerator's return as used in the 1851 Census

5.5.4 Availability of data

The inclusion of individual names on the completed forms raised the whole sensitive question of confidentiality. Unless individuals could feel absolutely certain that the information they gave to the enumerators would never under any circumstances be revealed to any other government department they would be unlikely to co-operate and so prejudice the results of the whole operation. In fact successive governments have extended confidence by retaining the 100 year rule whereby the actual returns made by the enumerators, and bound into books, are not made available to anyone outside the census officials for 100 years. Under this rule the 1871 enumerators' books are currently available. All the

No. of District	No. of Subdist.	SUBDISTRICT.	No. of Parish	Parish, Township, or Place	Area in Statute Acres.	HOUSES 1841 Inhabited	1841 Uninhabited	1841 Building	1851 Inhabited	1851 Uninhabited	1851 Building	POPULATION PERSONS 1801	1811	1821	1831	1841	1851
				327 KEYNSHAM.													
327	1	BITTON - - -	1	Kelston - - - - Parish	1095	47	2	–	45	3	–	221	248	248	274	255	259
			2	Northstoke - - - Parish	778	37	2	–	38	–	–	108	113	129	128	173	194
				Bitton (Glouc.), part of Parish—†													
			3	Bitton	3355	481	29	2	499	20	4	1094	1486	1788	2258	2413	2395
			4	Hanham - - - Chapelry	1212	258	16	3	261	31	–	795	934	1086	1212	1217	1180
			5	Siston (Glouc.) - - *uu* Parish	1827	196	1	–	201	16	–	856	833	902	973	1014	926
	2	OLDLAND - - -	1	Mangotsfield (Glouc.) - Parish	2591	820	39	7	871	45	2	2492	2901	3179	3508	3862	3967
			2	Oldland (Glouc.) - *x* Hamlet (part of *Bitton* Parish †)	2589	1205	56	10	1181	69	3	3103	3641	4297	5233	5708	5877
	3	NEWTON - - -	1	Saltford - - - *y* Parish	880	76	4	–	80	3	–	223	249	327	380	427	417
			2	Corston - - - *yy* Parish	1190	115	3	1	115	3	2	268	278	368	433	604	531
			3	Newton St. Loe - *yy* Parish	1578	100	1	–	88	4	1	371	384	431	477	527	440
			4	Priston - - - *z* Parish	1850	65	2	–	64	3	–	314	318	286	308	322	308
			5	Stanton Prior - - - Parish	841	29	–	–	32	1	–	131	144	158	159	148	149
			6	Marksbury - ᵥ - *aa* Parish	1277	67	–	–	68	2	–	283	306	354	371	328	310
			7	Compton Dando - *bb* Parish	1974	74	1	1	74	2	–	330	346	344	382	359	384
			8	Burnett - - - - Parish	608	18	–	–	18	–	–	64	70	75	82	100	95
	4	KEYNSHAM - - -	1	Keynsham (w) - *cc* Parish	4171	424	19	5	440	16	–	1591	1748	1761	2142	2307	2318
			2	Queen Charlton - - Parish	955	33	1	–	34	–	–	143	149	147	168	190	177
			3	Whitchurch, or Felton - Parish	2194	80	4	1	78	1	1	362	310	403	423	416	428
			4	Brislington - - - *dd* Parish	2393	214	2	3	213	13	–	776	1052	1216	1294	1338	1260
				Entire Parishes of													
		* Walcot - - - - -			1023	3853	416	11	4004	315	7	17559	20560	24046	26023	26210	27471
		† Bitton - - - - -			7156	1944	101	15	1941	120	7	4992	6061	7171	8703	9338	9452

Figure 5.12 A page from the printed volume of the 1851 Census Report

| POPULATION. | | | | | | | | | | | | HUNDRED. | NOTES. |
| MALES. | | | | | | FEMALES. | | | | | | | |
1801	1811	1821	1831	1841	1851	1801	1811	1821	1831	1841	1851		
												Bath Forum.	
120	132	108	135	128	135	101	116	140	139	127	124		
58	62	68	70	89	110	50	51	61	58	84	84		
541	731	883	1147	1238	1177	553	755	905	1111	1175	1218	Langley (Upper) (*Glouc.*)	
392	435	511	614	602	572	403	499	575	598	615	608		
420	396	462	466	469	447	436	437	440	507	545	479	Puckle Church (*Glouc.*)	
1259	1435	1540	1691	1831	1914	1233	1466	1639	1817	2031	2053	Barton Regis (*Glouc.*) Langley (Upper) (*Glouc.*)	
1490	1753	2069	2531	2786	2886	1613	1888	2228	2702	2922	2991		
113	125	178	192	207	212	110	124	149	188	220	205	**Keynsham.**	
126	139	186	218	320	277	142	139	182	215	284	254	Wellow.	
186	192	214	229	269	229	185	192	217	248	258	211		
149	154	146	150	160	154	165	164	140	158	162	154		
59	66	75	79	74	68	72	78	83	80	74	81		
141	162	194	186	164	162	142	144	160	185	164	148		
156	172	174	189	185	189	174	174	170	193	174	195		
28	34	37	45	51	49	36	36	38	37	49	46	Keynsham.	
783	825	870	1070	1128	1098	808	923	891	1072	1179	1220		
70	78	66	77	91	89	73	71	81	91	99	88		
191	153	182	197	194	214	171	157	221	226	222	214		
358	510	573	605	654	573	418	542	643	689	684	687		
												Entire Parishes.	
6829	7744	9541	10227	10442	10499	10730	12816	14505	15796	15768	16972	Walcot.	
2423	2919	3463	4292	4626	4635	2569	3142	3708	4411	4712	4817	Bitton.	

NOTES.

1841–51—Partis College is situate in this Parish, and contained 59 persons in 1841, and 62 in 1851.

327 KEYNSHAM.

u 327 ; 1 ; 3. The other part of Bitton Parish is in Oldland Subdistrict (327 ; 2.)

uu 327 ; 1 ; 5. 1851—The decrease of population in Siston is ascribed to the depressed state of the spelter works.

x 327 ; 2 ; 2. Oldland Hamlet includes the Village of Kingswood. The Parish of Bitton is chiefly in Bitton Subdistrict (327 ; 1.)

y 327 ; 3 ; 1. 1821—The increase of population in Saltford is ascribed to its vicinity to Bath.

yy 327 ; 3 ; 2—3. 1851—The decrease of population, in Corston and Newton St. Loe, is ascribed to the removal, since 1841, of large numbers of workmen who were then engaged in the formation of a railway.

z 327 ; 3 ; 4. Priston includes Wilmington Hamlet.

aa 327 ; 3 ; 6. Marksbury includes Houndstreet Hamlet. 1851—Several persons have emigrated from this Parish since 1841.

bb 327 ; 3 ; 7. Compton-Dando includes the Hamlet of Tucking Mills, and part of the Hamlet of Woolard.

cc 327 ; 4 ; 1. The Parish of Keynsham includes the Hamlet of Chewton Keynsham. 1831—The establishment of flax-manufactories in the Parish is the cause assigned for an increase in the population. 1841–51—The Union Workhouse is situate in this Parish, and contained 136 persons in 1841, and 153 in 1851.

dd 327 ; 4 ; 4. Brislington House Lunatic Asylum is situate in the Parish of Brislington, and contained 141 persons in 1841, and 139 in 1851. 1851—Large numbers of labourers engaged, in 1841, in the formation of the Great Western Railway, have since removed from this Parish.

enumerators' books have been transferred to the Public Records Office where they may be inspected in a special search room for census material. Enquiries for a reader's ticket that gives access should be addressed to The Keeper, Public Record Office, Chancery Lane, London, WC 2.

In Scotland the situation is slightly different. The census records are kept at Register House, Edinburgh and the enumerator's books are available at present up to and including the 1891 census. However and unfortunately from the researcher's point of view, Scottish practise is now being brought into line with English, and no further Scottish enumerators' books will be opened until 2001 when those for the 1901 census will become available.

Many local libraries in the larger towns will have transcriptions, Xerox copies or microfilm of their local returns. The 1841 enumerators' returns were in pencil and are not always very clear but those for 1851 are usually good clear entries in accomplished copperplate writing and relatively easy to work with.

There may be some advantage in transcribing the returns, having a separate slip for each family unit, using a form similar to that shown in *Figure 5.13*. When the details have been transposed it is then possible to break down the information and to reassemble it in a variety

| | | | | | PAGE: | AREA | | |
House No.	Name of Street, Road or House	Name & Surname of each person in house	Relation to Head of Family	Condition	Age of Male Female	Rank, Profession or Occupation	Where Born	If Blind Deaf Dumb

Figure 5.13 Transcription slip that can be used with enumerators' returns

of ways. The constitution of families and of households, the average age of males, females, and children under 16, for example, may be worth calculating, especially where two or more areas are being compared.

Occupations and trades can be analysed fairly easily though the change of occupational categories from one census to another after 1851 is a constant source of irritation since it makes it virtually impossible to study occupational changes over any long period. The number of servants and the number of families that employed servants is another subject worth pursuing. The column giving birthplace is also a guide to the degree of mobility of the population.

The 1851 Census Report gives comparable data for 1841 and at the same time gives the total population and the numbers of males and females for each of the preceding censuses.

Each census report has tended to become more elaborate than its predecessor but that of 1851 is probably the best with which to start off any fairly detailed local study, simply because it was the first to use full household schedules.

REFERENCES

Chibnall, A.C. (1973). *The Certificate of Musters for Buckingham in 1522*. Royal Commission on Historical Manuscripts. J.P. 18. London; HMSO

Drake, M. (1974). *Historical Demography, Problems and Prospects,* pp. 50—63. Milton Keynes; Open University Press

Eversley, D.E.C. (1965). 'Population in Worcester 1660—1850.' In *Population in History*. Ed. D.V. Glass and D.E.C. Eversley. London; Edward Arnold

Glass, D.V. (1973). *Numbering the People*. Farnborough, Hants.; Saxon House

Interdepartmental Committee on Social and Economic Research (1951). Guides to Official Sources. *No. 2. Census Reports of Great Britain*. London; HMSO. Now out of print but available in larger libraries

Laslett, P. (1977). *Family Life and Illicit Love in Earlier Generations*. Cambridge Group for the History of Population and Social Structure

Patten, J. (1971). 'The Hearth Taxes 1662—1689.' *Local Population Studies*, No. 7, pp 14—27

Peel, R.F.W. (1941). 'Local intermarriage and the stability of rural population in the English Midlands.' *Geography* **27,** 22

Sammons, P. (1976). 'Historical demography and the population experience of St. Mary, Harrow-on-the-Hill, 1560—1640.' Unpublished undergraduate dissertation, University of Bristol Department of Geography

Tate, W.E. (1946). *The Parish Chest*. Cambridge University Press

West, J. (1962). *Village Records*. London, Macmillan. An invaluable source for local study. On pp 134—135 there is a list of published and unpublished Hearth Tax Returns for different counties

CHAPTER 6 DIRECTORIES, RATE BOOKS AND NEWSPAPERS

6.1 INTRODUCTION

None of the themes dealt with in the different chapters is wholly unrelated to the others, and in this section especially there is a great deal of overlap with the other sections.

However the main emphasis is on urban rather than rural matters and the time-span extends as far back as the middle of the last century.

The most important source is the directories, because not only do they cover a long time-span in great detail, but they are also reasonably easily available and easy to work with. Used with care they open up much new ground and invite research on a variety of scales, ranging from the very small (changes in a single street over time for example) to the regional or national scale. One can, within limits, decide how much time is available and pitch one's investigation accordingly.

The rate books go back in some instances as far as the directories, but for many places they have been destroyed. The level of detail is very high indeed and they demand on the whole more effort for a given result than the directories.

Before starting on a project connected with any of these sources however one must be certain of one's objective. It may be a very limited objective indeed, for example to answer the question, 'What changes in the type of occupation took place in this street between 1860 and 1880?' The answer is likely to be more interesting if the changes were from private housing to shops, or from one type of shop to another. The longer the period covered by the study the more pronounced the changes are likely to have been.

In general it is more satisfying to put the question in the form of a hypothesis, and to seek the data in a form that allows one to test the hypothesis. The only problem is that some information is needed before one can put forward a hypothesis, and the hypothesis needs to be related in some way to the data that are available to test it. So it is always a valuable experience to start work on a new source, such as directories, with a limited objective, sufficient to make one familiar with the source, to allow one to find one's way confidently through it, and to be aware of what questions it cannot answer as well as what questions it can.

6.2 DIRECTORIES

6.2.1 Introduction

The many trade and commercial directories mainly of the nineteenth century provide one of the biggest potential and still largely untapped data sources for geographers, and they offer great scope for interesting and useful work at a variety of different scales.

The earliest directory in England was probably the Little London Directory, published in 1667. Thereafter directories appeared at an increasing rate. At first they were confined to London, Bath and other fashionable centres and were also rather selective in the areas covered and the kind of entries made. *Guide to the National and Provincial Directories of England and Wales published before 1865* (Norton, 1950) and *The London Directories, 1677—1855* (Goss, 1932) should be consulted before attempting work on any area before the mid-nineteenth century, and though these early directories offer less than complete coverage, they are the only available sources in many cases.

From the mid-nineteenth century the numbers and quality of the directories increased markedly, reflecting no doubt the growth of towns and the expansion of trade and commerce as well as a great increase in literacy on the part of the average city and town dweller.

For London the best source by far is the Post Office Directory, published annually since 1800. F.F. Kelly (1803—1883) left the Post Office to found his own directory publishing company. In 1845 he produced a directory for *The Six Home Counties* and in the space of the next 20 years new directories appeared for most of the English counties. After 1940 Kelly's ceased publication of county directories though they still issue biennial directories for many areas based on large provincial towns. The Kelly's format was so successful that they frequently had to take their competitors to court over copyright infringements and often bought up less successful competitors.

The Kelly's directories were produced at first every five or six years giving parish by parish coverage for one or more counties in each volume. For the earlier directories this can lead to problems of matching data from editions of different dates where the study area, say, involves two counties whose respective directories were published at different dates.

6.2.2 Format

The information for each parish was often supplied in the first instance by the local clergyman.

Figure 6.1 shows a copy of a typical entry for a village. Each entry consists essentially of three sections. First, there is a description of the parish, notes on the church, local history and places of interest. Local industries are generally mentioned, as are markets and fairs, though there is great variation in the quality of the information. Second, there is a list of 'Private residents' basically those who were neither tradespeople nor craftsmen but people of means who lived in the village. In small villages these come first even if not so described.

Date Thomas, farmer, Beggearn huish
Davis Edward, blacksmith
Davis Edwin, plumber & glazier, Yard
Edbrook Ann (Mrs.), shopkeeper

Greed John, farmer, Holcombe water
Hill John, wheelwright, Yard
Hosegood George, farmer, Huish Barton
Lloyd John, farmer

Milton Henry Robert, miller, Yard mills
Prideaux Robert, farmer, Slade
Stone Thomas, farmer, Clitsome
Tapp Philip, farmer, Yard farm

NEWTON ST. LOE is a parish and beautiful village, 2 miles east from Twerton station, 3½ west from Bath, and 2 south-east from Saltford station, in the Eastern division of the county, eastern division of Wellow hundred, Keynsham union, Bath county court district, rural deanery and archdeaconry of Bath, and diocese of Bath and Wells, situated on the river Avon, which is here crossed by a stone bridge. The church of Holy Trinity is a handsome stone building : it has a chancel, nave, aisles, tower with 2 bells, and two porches : it was thoroughly restored and enlarged, and one aisle added, in 1857, at the expense of W. G. Langton, esq., and the chancel was at the same time rebuilt by the rector : all the windows are filled with stained glass : in the church are some handsome tablets to the Langton family ; and in the churchyard is a marble monument to Captain Edward Gore Langton. The register dates from the year [   ]. The living is a rectory, tithe rent-charge £410, with 48 acres of glebe and residence, in the gift of William Stephen Gore Langton, esq., and held by the Rev. George Gore, M.A., of Emmanuel College, Cambridge. There is a school for boys and girls, being children of poor inhabitants of Newton St. Loe and Stanton Prior (boys of Newton only being apprenticed), endowed with £72 a year, by will of Richard Jones, 1698. There is a school for boys and girls, from four years of age to seven, and a boys' Sunday school, both supported by the rector ; also a girls' Sunday school, supported by subscription. The charities are numerous, amounting to £78 yearly. Newton Park is the seat of William Stephen Gore Langton, esq., D.L., J.P., lord of the manor and chief landowner. The soil is stone brash ; subsoil, blue lias. The chief crops are wheat, barley and oats. The area is 1,578 acres ; rateable value, £6,765 ; and the population in 1871 was 417.

Parish Clerk, Charles Glover.

POST OFFICE.—George Mercer, postmaster. Letters from Saltford R.S.O. arrive at 8 a.m. ; dispatched at 7.15 p.m. The nearest money order office is at Twerton

SCHOOLS :—
Endowed Public Elementary, Charles Glover, master
Infant do. Miss T. Davies, mistress

Gore Rev. George, M.A. [rector], Rectory
Langton William Stephen Gore, D.L., J.P. Newton park
Maule Henry St. John, The Hayes
Milward Miss, Prospect cottage
Sprye Rev. Theodore de L. [curate]
Stothert John Lum, Mount view

Collins James & Son, millers & farmers, Newton mill ; & Dead mills, Lower Swanswick
Edwards Samuel, grocer & baker
Evans John, bailiff to W. Stephen Gore Langton, esq
Gray James, shoe maker
Hathway John, *Globe,* & farmer

Mercer George, blacksmith
Osborne Hen. farmer, Pennsylvania frm
Pawsey Edwin, farmer
Smallcombe Aaron, farmer, Clay's end
Willis John Gale, farmer & maltster, Newton farm & Manor farm, Wellow ; malt house, Corston
Wittington William, shopkeeper

NORTHOVER is a parish and village, adjoining Ilchester, in the Mid division of the county, Tintinhull hundred, Yeovil union and county court district, Ilchester rural deanery, Wells archdeaconry, and diocese of Bath and Wells, situated on the river Ivel. The church of St. Andrew, an old Gothic building, has a chancel, nave, north aisle, tower with 4 bells, and contains an organ. The register dates from the year 1600. The living is a vicarage, endowed with great tithes worth £140 yearly, exclusive of residence and 11 acres of glebe land, in the gift of Mrs. Burnard, and held by the Rev. Lyttleton H. Powys Maurice, M.A., of Caius College, Cambridge. Mrs. Burnard is lady of the manor and chief landowner. The soil is loamy ; the subsoil is gravel. The chief crops are wheat, beans, and barley. The area is 436 acres ; gross estimated rental, £1,230 ; rateable value, £1,143 ; and the population in 1871 was 90.

SOCK DENNIS, formerly extra-parochial, is now a parish, in Yeovil union : has an area of 880 acres, and a population of 28.

Parish Clerk, Samuel Warren.

Letters through Ilchester, the nearest money order office
INSURANCE AGENT.—*West of England & Railway Passengers,* J. Baker

Figure 6.1 Part of a page from the Post Office Directory of Somerset, 1875

Third, there is a list headed 'Commercial' arranged in alphabetical order by surname of those engaged in business. This is interpreted widely to include retailers, service and manufacturing, farming, rural crafts, and professionals such as doctors, dentists and veterinary surgeons.

It is fairly certain that these lists in the parish sections are reasonably accurate. Appended to the Commercial section are notes on the carrier, coach train and bus services to the principal towns nearby.

In larger towns and cities the Commercial section is arranged in a more systematic way with an alphabetical list of trades and services; and under each heading is a list of those who offer the service together with their address. *Figure 6.2* shows a page from the classified Commercial section for the City of York in 1861.

Another section of the directories is the 'Court Directory', an alphabetically arranged list of those private residents already recorded in the parish section. Since the criterion for including these people is not stated there is not much that can be done with this information. As much as anything else it is probably now best seen as a mute and rather sad comment on the desire of many in the Victorian period to mark their upward progress through the ranks of society. To those who were uncertain of their position, or else very well aware of it, the appearance of their name in this section may have been a source of reassurance.

A final section is an alphabetically ordered list of trades and services for the whole country, based on combining the individual entries. Tempting as it may be to use this conveniently arranged list, a word of caution is necessary, because in this section there is a great deal of double-, even of triple-listing. For example, a person who in a village entry was described as a blacksmith and carpenter might well appear under each of these headings in this final section. Or he might appear only once, either as a blacksmith or as a carpenter. In other cases a person listed in a village may be omitted entirely from this final classified section. So on the whole it is best wherever possible to avoid using this section unless one is prepared for a degree of inaccuracy that could in some instances be as high as 10%, but which could be lowered by careful cross-checking.

The type of map that can be prepared from the Post Office Directory for London and from the Kelly's or other similar directories for the larger provincial cities is well shown in *Figure 6.3* based on work carried out by Oliver (1964). It reveals very clearly the great variety of establishments supplying the furniture departments of the retail stores of Maple's

PUBLICANS—continued.
Star & Garter, E. Williamson, 3 Nessgate
Sun, W. Morrell, 4 Tanner row
Tam O'Shanter, C. Leoley, 19 Lawrence street
Thomas's hotel, W. Thomas, Museum street
Three Cranes, W. Briggs, 16 Sampson square
Three Cups, J. Smithson, Walmgate
Three Tuns, Maltby & Wilberforce, Coppergate
Tiger, J. Burrill, 15 Market street
Trumpet, J. Mintoft, Townend street
Turf tavern, J. Thompson, Market street
Turk's Head, R. Garnett, 31 St. Andrewgate
Turk's Head, J. Marsh, 17 College street
Turk's Head, J. Walker, 51 Great Shambles
Unicorn, Mrs. M. A. Rogers, 10 Tanner row
Victoria, J. Lambert, 1 Dove street
Victoria, R. Ramsdale, Cemetery road

RAG MERCHANT.
Hardcastle J. Lady Peckitt's yard

RAILWAY CARRIAGE AND WAGGON BUILDERS.
Knapton William & Charles, Albion foundry, Aldwark

READY MADE LINEN WAREHOUSE.
Outhwaite W. 18 Coney street

REFRESHMENT ROOMS.
Baren G. 28½ Parliament street
Fall G. 38 High Petergate
Foster J. Coppergate
Marston R. J. 27 Parliament street
Nowell T. 2 High Petergate
Petch T. 22½ High Petergate
Sowden Mrs. A. 18 Davygate
Stamper J. R. 6 Market street

ROPE, LINE AND TWINE MANUFACTURERS.
Clapham G. George street
Dalton J. 20 Great Shambles
Hessay J. 37 Great Shambles
Lawson John, 5 Micklegate
Lawson J. 6 Walmgate
Lund William David, 50 Low Petergate
Shephard G. 182 Walmgate
Walker J. 21 Lord Mayor's walk
Wardby Mrs. M. 40 Great Shambles

SACK MANUFACTURER.
Lund W. D. 50 Low Petergate

SADDLE TREE MANUFCTR.
Edson W. School lane

SADDLERS AND HARNESS MAKERS.
Atkinson J. 46 Micklegate
Clark J. 7 Margaret street
Cook W. Walmgate
Cooper M. 3 Railway street
Cutbert G. 44 Bootham
Eland J. W. 60 Walmgate
Frank J. 185 Walmgate
Gray G. 1 Fossgate
Holtby Mrs. H. 2 New Bridge street
Johnson R. 20 High Petergate
Kendall W. 10 Goodramgate
Mason & Co. 17 Spurriergate
Nutt M. 92 Goodramgate
Oxtoby B. 183 Walmgate
Smallwood R. 161 Walmgate
Wiggins D. 29 Coney street

SADDLERS' IRONMONGERS.
Ellis R. & Son, 25 Castlegate

SAILMAKERS.
Fisher J. King's staith
Pratt J. King's staith

SALT MERCHANTS.
Etty T. Old foundry yard
Gladin J. Friargate

SAND MERCHANT.
Tesseyman R. 33 North street

SAUCE MAKER.
Postill Edward, 7 High Petergate. See advertisement

SAW AND PLANING MILLS.
Bellerby J. &Son, Piccadilly, &George st
Condeaux & Ernest, Patrick's pool, Church street

SCHOOLS.
Marked thus * are for Girls.
Marked thus † are for Boys.
*Armstrong Mrs. Fulford field house, Fulford road
*Bedford C. Trinity lane
*Brooks Mrs. H. 5 Tower street
Coates W Priory street
*Dawson Mrs. M. 7 College street
*Eastburn Misses Dorothy & Mary, East Mount road
*Fenwick Miss A. 32 St. Mary's
*Gilbert Misses Mary & Charlotte, 36 St. Mary's
*Heselwood Miss M. E. 3 Ogleforth
*Husband Misses M. A. & J. Burton house, Clifton
*Hustler S. School lane, Navigation street
*Jackson Misses H. & L. 8 Bootham
*James Miss E. Coffont place
*Johnson Miss A. M. Middlethorpe hall, Bishopthorpe road
Keller J. 10 Church street
McClellan James Creighton (boarding), Clementhorpe house, Nunnery lane
*Matterson Misses (Jane & Fanny) & Appleby (Maria), 25 Bootham
*Mercer Miss M. A. Minster yard
*Monkman J. 11 Lord Mayor's walk
Mosley R. Holgate road
*Murray Miss M. 10 St. John's street
*Newton Miss C. 1 Castlegate
*Noble Miss E. 24 Holgate road
*Patterson W. T. 11 Pavement
*Pickersgill Miss M. A. 5 St. Saviour pl
*Quarton Miss M. 18 Colliergate
*Roberts Mrs. M. Heworth moor
*Sample Miss A. Jull lane
Scott R. George street
*Shorter H. Chapel row
*Stephenson Misses M. & E. Bishop hill senior
Stoker Miss M. A. 26 Mount Ephraim
*Tabor J. 5 Penley Grove street
*Thorp F. 20 Bootham
*Tregelles Miss R. the Mount
*Wilkinson Mrs. M. A. 16 South parade
*Wilson H. 8 Spurriergate
*Wright Misses J. & S. 29 Stonegate

SCULPTORS.
Bradley Robert, 19 Gillygate
Fisher Mary A. & Son, 22St. Saviourgate
Jackson W. E. 15 Coney street
Plows W. Walmgate
Skelton M. & Son, 81 Micklegate, & Tanner row

SEED MERCHANTS,
Horner G. 20 Little Stonegate
Lawson J. J. (elder), 5 Micklegate
Lund W. D. (elder), 50 Low Petergate
Rooke M. & Son, 13 New Petergate
Tonge J. S. 57 Low Petergate, & 55 Goodramgate

SEEDSMEN.
See Nurserymen.

SERVANTS' REGISTRY OFFICES.
Britton Mrs. M. 14 Little Blake street
Dixon Mrs. A. 10 Colliergate
Oates Miss C. Minstergates
Sigsworth Mr. & Mrs. 24 Coney street
Usher Mr. W. 5 Market street

Waggon & Horses, J. Dawson, 22 Gillygate
Waggon & Horses, R. Fisher, 11 Lawrence street
Wellington, L. Beeby, Goodramgate
Wheatsheaf, W. M. Briggs, 19 Davygate
Wheatsheaf, R. Scruton, 12 Hungate
White Hart, T. Beck, 26 Stonegate
White Horse, J. Cowper, 24 Skeldergate
White Horse, W. Wade, 60 Bootham
White Horse, G. Ward, Coppergate
White Storn, R. Wilson, 57 Goodramgate
White Storn, T. Foreman, Pavement
Windmill, W. Hunter, 7 Blossom street
Woolpack, T. Bowman, St. Saviour place
York Arms, G. Mitchell, 94 High Petergate
York Glassmakers' Arms, J. Willans, Cattle market
Yorkshire Husser, G. Bedall, 10 North street
Yorkshiremen inn, Thomas Buck, Coppergate

Ward Mrs. M. 11 Stonegate
Ware Mrs. H. 54 Stonegate
Wilkinson W. 8 Rougier street

SHIPBUILDER.
Horner J. M. Clementhorpe

SHIPOWNERS.
Green G. Queen's staith
Leetham J. J. 27 Skeldergate
Senior L. Wesley place

SHIRT MAKER.
Milward C. A. 20 Coney street

SHOEMAKERS.
See Boot & Shoe Makers.

SHOPKEEPERS.
Allanson Mrs. J. 48 Skeldergate
Appleby G. M. 12 Walmgate
Armitage G. 37 North street
Arnett T. 10 Tanner row
Atkinson C. Heslington road
Atkinson R. B. Blossom street
Atkinson T. Townend street
Bagnall J. Hope street
Bainbridge Mrs. M. 63 Goodramgate
Barker J. 8 King street
Barker J. 14 North street
Barker J. Powell pl. Lord Mayor's walk
Barnes S. Middle Water lane
Bawtry Mrs. M. 103 High Petergate
Beck J. Vine street
Beckett J. 36 Oxford street
Bell W. Hope street
Bellerby J. 15 Swann street
Benson J. 31 Gillygate
Birch J. Clementhorpe
Birch W. Peaseholme green
Bland H. Bilton street
Bousfield J. Margaret street
Bowman J. 12 Aldwark
Brooke J. 68 Walmgate
Brooks W. 7 Lawrence row
Broom Mrs. A. Lowther street
Brown A. 54 Micklegate
Brown A. Alma terrace
Brown M. Hope street
Brown R. 3 College street
Brown T. 04 New Biggin street
Browne Mrs. A. 8 North street
Brownlee T. Lord Mayor's walk
Bulmer T. 34 Layerthorpe
Burrill J. 6 Fishergate
Byers W. Long close lane
Calam Miss M. Toft green
Camage Miss E. 19 George street
Cappelman Mrs. S. 28 Penley Grove st
Carlutt Miss E. 22 Grape lane
Carbutt Mrs. M. Marygate
Cawood Mrs. E. Cemetery road
Chapman J. 7 Fetter lane
Clark R. 103 Walmgate
Cloak E. 14 Layerthorpe
Cole J. Townend street
Cooper R. Nunnery lane
Covell W. 12 Goodramgate
Cowper E. George street
Craven J. 116 Walmgate
Croft W. 108 Walmgate
Cutherl J. 93 Walmgate
Dalby Mrs. H. 18 North street
Dalby Mrs. S. Middle Water lane
Dawson T. 107 High Petergate
Dearlove Mrs. A. 5 Sampson square

Figure 6.2 A page from the Post Office Directory for the City of York, 1861

Figure 6.3 Linked furniture-making establishments, Tottenham Court Road and Charlotte Street, 1872. 1. Cabinet maker, fancy cabinet maker, cabinet liner, cabinet inlayer; 2. buhl cutter; 3. cabinet carver; 4. french polisher, japanner; 5. turner; 6. varnish maker; 7. chair maker; 8. table maker, billiard table maker; 9. upholsterer; 10. upholsterer and cabinet maker; 11. leather merchant; 12. cane worker; 13. upholstery spring maker; 14. upholsterers' warehouseman; 15. bedstead maker; 16. bedding manufacturer; 17. carpet warehouse; 18. timber merchant, mahogany and rosewood merchant, veneer merchant; 19. saw mill; 20. carver and gilder; 21. looking-glass maker; 22. wholesale ironmonger; 23. tool maker; 24. salesman, furniture broker. (After Oliver (1964) *Geography*, p. 403, courtesy of The Geographical Association)

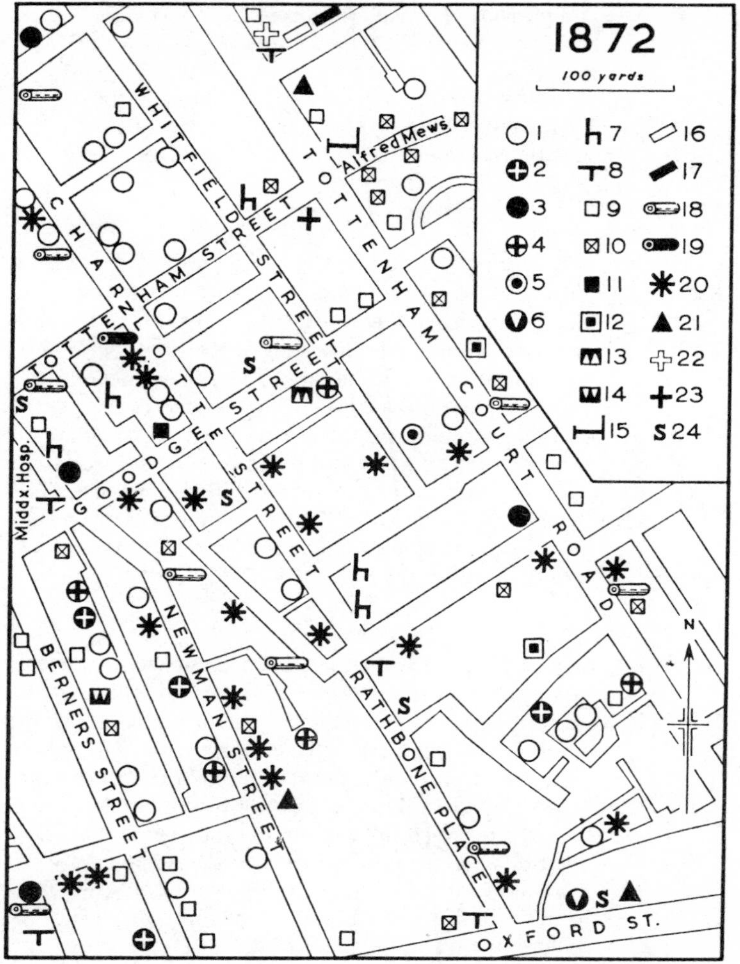

and Heal's and others in Tottenham Court Road in 1872. There is no other source as accessible for such detailed material in this period, and there is still much scope for investigation of urban industries.

In the case of a city or town the entry in the early years normally takes the same form as for the village though later there is a street-by-street section which should be used wherever possible in preference to the classified trades section at the end of the directory.

6.2.3 Problems

For most purposes the Kelly's directories can be used with confidence. Were there too many errors the directory would not have been commercially viable — people would not pay for an inaccurate product, and inaccuracies would tend to be reported and the corrections incorporated in later editions. But the time that their publication spanned was one of great, and in many cases, quite rapid change and there are bound to be some inaccuracies. This is also true of data that one collects oneself in the field, however careful one is to keep the errors to a minimum, and it is no reason for not using the directories. The main errors seem from experience to be in the classified trades section so it is sensible wherever possible to work from the entries under individual villages and towns.

Most large reference libraries should have or be able to obtain a collection of the Kelly's directories for at least the local area or in some cases they will have a local equivalent that is similar in format.

6.2.4 Case studies

Case study 1 — Changes in local accessibility in Devon and Cornwall 1856—1939
The aim of this study was to discover the effects of changes in accessibility between villages and towns on the system of central places. Major changes in accessibility, in space relationships, came with the growth of the railway system and with the introduction of the motor bus that replaced both the old horse-drawn carriers' vans and the faster mail coaches. Directories were the only source of information on the services offered by the carriers who travelled between the villages and the market towns of the area. A village carrier is usually mentioned twice, once under the name of the village from which he operates and once under the main town which is his destination. Confusion sometimes arises when a carrier leaves from his home village A and passes through villages B,C,D and E before reaching town F because the directories sometimes enter the carrier under each of the villages he serves en

route without making it clear that he is in fact merely passing through most of them. Another problem arises if a list of carriers is compiled only by looking under the departures from the main market town. Often a number of small carriers from distant villages are left out of the market town entry.

Figure 6.4 shows the pattern of regular carriers operating to and from the market town of Truro in 1856. By 1914 the motor bus was beginning to replace the horse-drawn carriers' wagons and brakes and Kelly's usually give details of the bus services and the timetables for each village. The period after 1914 was the heyday of the private bus and coach proprietor, with dozens just in Devon and Cornwall offering a more frequent and varied service than rural areas were ever to see again. By 1939 many of the independent operators had been absorbed by the large national groupings and the range and frequency of services in many areas decreased. Using the information from Kelly's, supplemented in the later period by published timetables issued by the bus companies, one can reconstruct the changing area served by each of the principal towns in Devon and Cornwall.

Figure 6.5 shows what are called 'core areas' for each of the major centres; these are areas over which the respective centres have exercised undisputed sway between 1856 and 1939. Where an area is served by two centres it lies outside the core area of each. The map in effect shows stability and instability. Where most of an area is covered by core areas, with only narrow zones of overlap between them, we can infer that the relationship between town and village has been fairly stable in the period 1856—1939. Where there are large zones separating adjacent core areas then there has either been constant competition throughout or a change in the relationship between the villages and their respective local centres.

One aspect of the study was the changes in spatial relationships between villages and centres. The other was changes in the system of central places themselves as revealed by the changes in the kind of services they provided. These changes can be identified by using the information in the directories. In this particular study a selection of important services was made, taking care as far as possible to choose only those services likely to be as common or as needed in the early as in the late part of the period considered. For example, it would not have been sensible to have included garages or photographic dealers or cinemas in the key list of services since they would not have been available in 1856.

The selection of the services to be considered is rather a critical matter, but in the interests of brevity it will be passed over at this stage. The list of key services having been decided on, each settlement in turn was given a points score based on the services it offered in each of five periods: 1856, 1883, 1914, 1923 and 1939 (these correspond to the periods

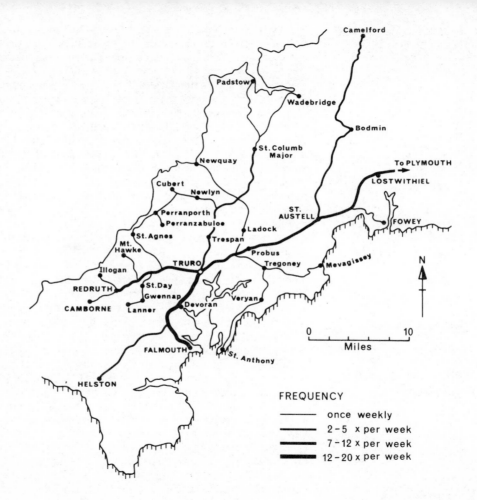

Figure 6.4 Carriers serving Truro in 1856 (From Kelly's Directory of Cornwall, 1856)

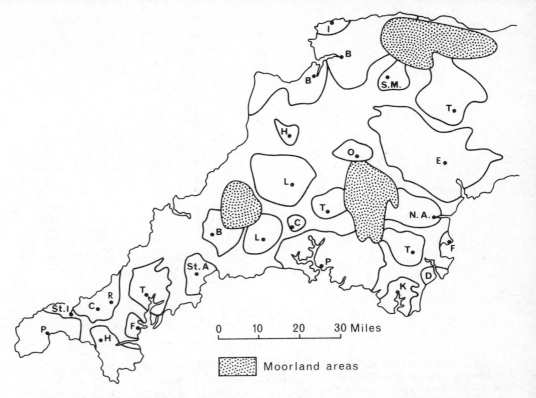

Figure 6.5 Core areas in Devon and Cornwall 1856—1939. Capital letters identify the principal market towns

Moorland areas

for which the accessibility maps were prepared). The higher the score assigned to a particular settlement the greater the *range* of services provided. The number of establishments providing each service was not considered in the final score, only the presence or absence of the selected functions. It was possible to do it this way because there was a hierarchy of functions in all settlements. Certain very specialised functions were only found in the more important places and where these were found there was also a full range of the more common functions.

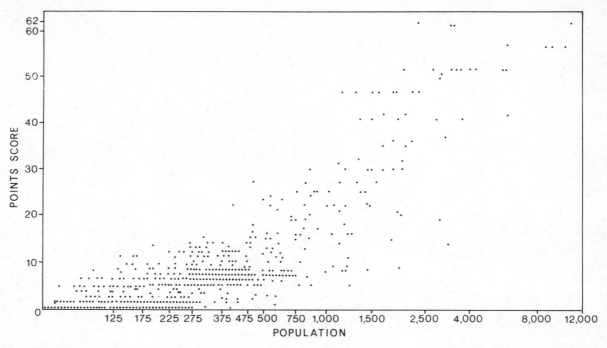

Figure 6.6 Devon and Cornwall 1856. Relationship between population and services as indicated by points scores

Then the score for each settlement for each of the five periods was related to the population of that settlement. Not unexpectedly there proved to be a broad relationship between the range of services provided as shown by the points score and the population of the settlement (*Figure 6.6*) However, the relationship was neither simple nor consistent and places with the same population often had widely differing numbers of services associated with them. So it was then necessary to identify places that had an unusually *high* level of services associated with them for their size in each of the five periods. When the places with a relatively high score for each period were plotted it was clear that they tended to be either in the centre of core areas, usually in established market towns, or, more interestingly, on the margins of core areas. This is clearly shown in *Figure 6.7.* In 1856, with

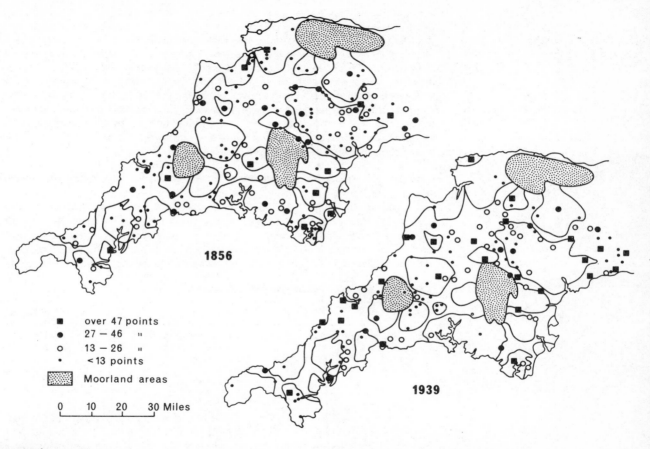

over 47 points
27 — 46 "
○ **13 — 26** "
· **<13 points**

Moorland areas

0   10   20   30 Miles

1856

1939

**Figure 6.7 Centres and core areas
1856 and 1939**

accessibility generally poor and relatively little movement, there were a fairly large number of centres with relatively high scores, many between core areas and others within core areas. By 1939 with greater accessibility the 'high' centres within core areas had largely disappeared except in the largest core area (Exeter) where distance still protected them from competition. This change in fact can be pinpointed to the period when the motor bus and later the private car greatly increased personal mobility and people were able to travel further afield on shopping and entertainment trips, thus favouring the larger if more distant centres. Only when remote from the influence of the large centres were many of the smaller centres still able to thrive. Even today, more than half a century after the introduction of modern buses, some of the smaller centres in Central Devon and North Cornwall support a higher range of services than one would expect.

Case study 2 — Locational behaviour of urban retailing in Hull during the nineteenth century
In this study by Wild and Shaw (1974), the basic information came from three trade directories of 1823, 1851 and 1881 which covered the period after fixed shops had become common in most towns and large cities but before retailing patterns were affected by changes in urban transport systems, particularly electric trams in the 1890s and 1900s. Maps of the city contemporary with each directory were available. The retail trades selected are shown in *Table 6.1*. The method of recording the data was based on grid squares, each

TABLE 6.1 RETAIL TRADE GROUPINGS IN HULL

Clothing trades	Other trades	Food trades
Milliners and dressmakers	Chemists and druggists	Grocers
Tailors, drapers and clothes dealers	Booksellers and stationers	Butchers, ham and cheese dealers
Boot and shoe makers	Pawnbrokers	Confectioners
	Jewellers	
	Ironmongers	

square measuring 200 yd by 200 yd (3.35 ha). The first square, that determined the position of the others, was placed on the peak land value point for the city as it was in mid-century. From each square were subtracted all areas not given over to residential or commercial uses, leaving a net residential and commercial area for each square. Then for each period a percentage (*P*) representing net/gross area was calculated for each square. Only squares

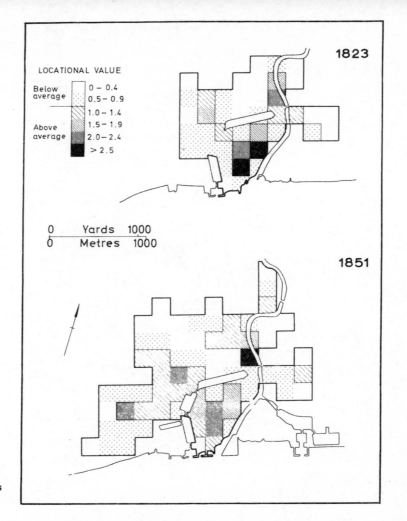

Figure 6.8 Composite location of shops in 1823 and 1851, Hull

scoring 25% or more were counted as part of the built-up area. The number of shops per square was then recorded and the Locational Value (*L*) for each square was calculated from the following formula:

$$L = \left(X \times \frac{100}{P} \right) \div \frac{N}{A}$$

When *N* = total recorded shops and *A* = net area of town.
The shading used in *Figures 6.8* and *6.9* indicates degrees of variation from an average situation, represented as a locational value of 1.0.

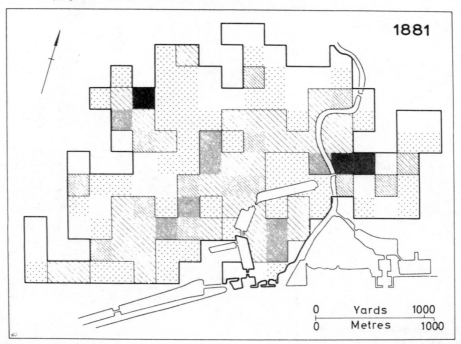

Figure 6.9 Composite location of shops in 1881, Hull

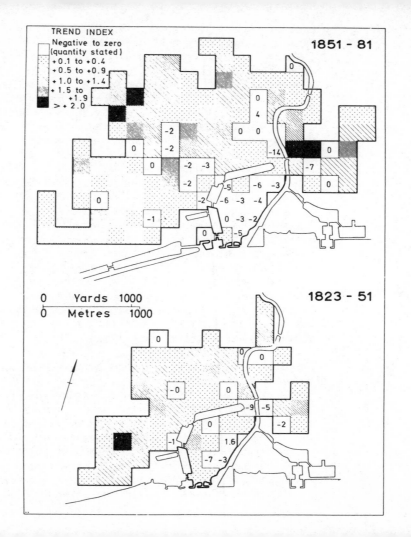

Figure 6.10 Trends in shop location, Hull

Once the locational value has been established for each square and for each of the three dates, it is possible to see how much change there has been between any two periods (*Figure 6.10*). From the two sets of maps and from a detailed analysis of what was happening to different groups of shops, certain quite clear conclusions can be drawn.

In 1823 58.5% of the city's shops were in the historic Old Town and only two districts in the suburbs had above average locational values. By 1851 the proportion of total shops in the central Old Town area had fallen to 38.9% though the actual numbers had increased, but only slowly, and there had been a relative shift of shops to suburban locations. In fact the fastest growing areas of new shops corresponded to the major areas of working class housing. The squares with below average values picked out the middle class areas. Between 1851 and 1881 there was an overall 76.3% growth in the net residential and commercial area of the town, and increasing suburbanisation of shopping, the growth now being mainly in the outer suburban areas.

As the district developed with working class housing it seems that there was an initial colonisation by the relatively dispersed trades, such as milliners and dressmakers, drapers and tailors, butchers, boot- and shoemakers, grocers and general shopkeepers. The more specialised shops remained relatively concentrated in the Old Town and their customers tended to be mainly drawn from the dispersed but more mobile middle classes who were able to take advantage of the facilities in the Old Town. Mature suburban centres attracted to themselves a proportion of the more specialised shops in due course. The findings of Wild and Shaw in Hull may or may not be echoed by future studies along the same lines, but there is no doubt that the directories in the hands of experienced and careful workers can shed a great deal of light on a neglected area of urban retailing development.

Case Study 3 — The changing numbers and location of pawnbrokers in Bristol, 1890–1940 (Figure 6.11).

The initial locations as revealed by the 1890 directory were mainly in the central areas of the city, reflecting the compact size of the city which in turn was a function of the small population and the limitations of the tramway system. The 1890 directory describes their location as being 'so arranged that a person may call upon the whole of them in three walks without going twice over the same ground. This arrangement will be found useful to those who may have lost plate, etc.' This comment shows a nice regard for the sensibilities of the readers of the directory; most of those who would have used the pawnbrokers were poor people who pawned possessions for cash, in the hope of redeeming them later when

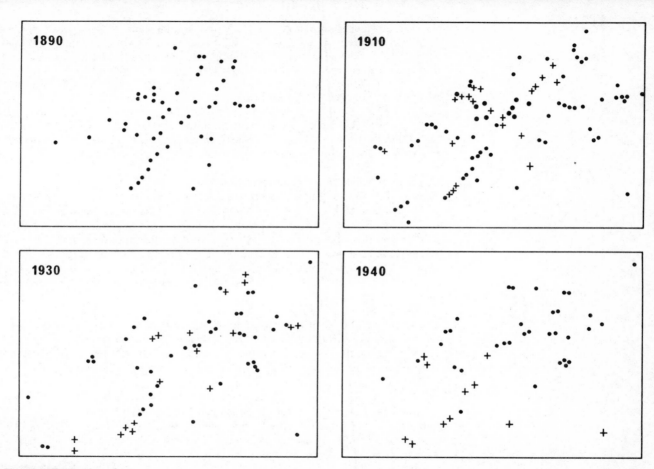

Figure 6.11 Changing distribution of
pawnbrokers in Bristol, 1890—1940,
from local directories

their circumstances improved. Clustering was natural where potential customers were found in certain areas only and where they might be expected to shop around for the best terms.

By 1910 some had closed but even more had opened, with the areas to the southwest, east and northeast gaining significance. It was here that there had been a considerable amount of building of working class houses, near to the industry in suburbs such as Bedminster, Lawrence Hill, Easton and Barton Hill.

By 1930 the peak had passed and there were many more closures (18) than openings (4) in the previous decade. 1940 saw the numbers further reduced, to a total of 34 compared with 69 at the peak in 1910, and the former concentration on the central area had given way to a general dispersion towards the mainly working class suburbs to the east.

Directories and studies in residential mobility

Figure 6.12 shows a typical page from a directory for a large town in the nineteenth century. In fact this comes from the Wright's Directory for Bristol in 1878. Each house is listed and for each house there is one or more names. One has to presume that the name given is that of the head of the household; it may or may not be that this person owns the house. Where two different names are given the assumption is that the accommodation is shared in some fashion. Occasionally some indication is given of the occupation or profession of the occupant, presumably at his discretion unless he occupies premises from which he also carries on his trade or profession.

A new edition of the directory was normally produced each year and it would need to be reasonably accurate or it would not be a commercial proposition. So the question arises, can we use the information given in the directories to gain some idea of the amount of residential mobility of the population in the past, to calculate the frequency with which certain properties, streets or districts changed occupants at different periods?

The rate books, as we shall see later, are a better source for this purpose since they were more accurate and were in most cases revised every three months. The shorter interval between rate books means that they are much more likely to capture short term residents, and many in Victorian times took rented accommodation on a quarterly basis. But rate books are not always available, often having been destroyed or mislaid in the past and in such cases the directories are one of the few alternatives left.

Figure 6.13 shows a suitable format for recording the result of a pilot investigation into residential mobility using the directories. The simplicity of the example conceals many irritating and sometimes insurmountable difficulties. For example, as the Victorian city

Sidney Place, *Redland.*

Supply Co.
1 David Davies, *Wellington Arms*
2 Abel Stock
3 Sidney White, boot maker
4 Mrs Smith
5 James J. Case, builder
6 T. Haddon Smith, Whitson villa
   Mrs Heather, Coldson villa

Sidney Row, *Cumberland Road.*

1 Patrick Culleton
   June Ann Propert, grocer

Silver Street.

1 Richd Stinchcombe, provision dler
2 John Pollard, pork butcher

Sims's Alley.

here cross over

13, 14 Mrs Austin, clothes dealer
10 M. Mapstone, news agent
11 S. Bendall, *Berkire*
11, 12 Jas. Albert Whittand, clothes dlr.

here cross over

1 James Pyne, straw cleaner & dyer
2 F. H. Pearson, ticket writer
4 Alfred Barber, jeweler
5 F. and R. Deacon, hatters
7 Robert Dove, clothes dealer
8 Jefferson and Co., hatters
9, 10 William Wotman, confectioner

Sion Hill, *Clifton Down.* —manager, Mr. Loyatt

St. Vincent's Rocks Hotel —manager, Mr. Loyatt
2 Miss Strange, lodging house
3 Ruth Jenkins, lodging house
4 Tom Partridge, lodging house
5 John Scadding, lodging house
6 George Wildbore, lodging house

Sion Lane intersects.

7 George Wildbore, lodging house
   Miss Finlay
9 Power's Boarding House
9 Richard H. Walwyn
10 Miss Yeend
11 George Robert Hall, lodging house
12 John Wilson, lodging house
13 Mrs Bragge, lodging house
14 Mrs Butler, lodging house
15 Henry Pearpe, stationer, baths
16 Andrew Woodward, lodging house
17 Mrs Mary Sheard, lodging house
   Mrs M. Ferguson
18 Miss Edgar
18 Coles & Redfern, lodging house
19 Miss Conway
   Charles Morley, lodging house
   Miss Boraston
   Miss Wantall
   Miss Parkes
20 George Coyle, lodging house
21 Miss Gibbons & H. Burroughs
   Mrs Stead, lodging house
24 Frank Weaver, surgeon dentist
   James Blake, lodging house
26 James Harris
   Mrs Leal, lodging house
27 Samuel Carey, lodging house
   The Misses Nosterman
   Mrs Pavey, lodging house
30 Elizabeth Chapman, fly proprietor
   Clifton Down Hotel—Daniel Gittins, manager

Sion Mews, *Clifton Down.*

1 Elizabeth Clarke, *Clifton Down Hotel tap*
   John Bryant, coach builder
2 Mary Davis, *Suspension Bridge Inn*
3 Mrs Skeel

4 William Harris
5 George Brown, *Coronation tap*
   Elizabeth Chapman, fly proprietor

Sion Road, *Stapleton Road.*

John Watson, brush maker
Voisey and Rogers, leather dressers
J. Edwards, bootmaker
Thomas Newland, general dealer
Mrs Kingston, general dealer
John Stephens
George Marley, dairyman
William Nutt, bootmaker
John Cooper, tailor
Isaac Upton
Miss Mountain, laundress

Norfolk Place.

1 Joseph Rogers
2 Miss Phillips
3 Frederick Jenkins

Norfolk Buildings.

E. S. Colman, licensed malt-ster

Small Street.

New Buildings—

Luke Arnold, stock broker
Charles Bevan and Charles Robt
   Hancock, solicitors
Wadham, Chilton, and Green-
   Armytage, solicitors
Ryland & Cozens, public and
   private accountants
J. T. Broad, solicitor
Samuel Pedden & Co.
Wm. Rice
Bristol Times and Mirror, Leech
   and Taylor
Post Office, Money Order Office,
   and Telegraph Office—Post-
   master, E. C. Sampson
Bristol Water Works Co.—Sec,
   Alfred J. Alexander
7 Bartlett and Mogg, wine merchants
   dealers
8 Clement Gardiner, accountant, etc
8 J. H. Hirst, architect and surveyor
9 H. N. Sleigh

here cross over

St. Werburgh's Chambers—

County Court Offices
Edward Harley, solicitor and
   registrar of County Court
   Edwd. Arthur Harley, solicitor
   and registrar of County Court
9 Charles T. Lloyd & Co, linen mchts
   Thos. Holmes & Co., wine merht.
10 Lewis Brothers, cheese factors
11 G. C. Ashmead and Son, surveyors
   Jonathan Perrin, solicitor
   Edward Thomas, solicitor

here cross over

Crown Building Society
12 Arthur Jeffery Morris, wine merht
   Henry Brittan, Press and Inskip,
   solicitors
12 Edw. Wilkins, carpenter & undertkr
13 James Wright, cheese & prov. fctor
13 Colston Building Society, Geo.
   Morris, secretary
14 Whittington, Grible, and Gould-
   smith, solicitors
   James T. Tilley, stock and share
   broker, Secretary to The Queen
   Building Society
15 *Justice Courts Hotel,* Alfd. E. Gould
15 Beaufort Club

Fry's Buildings—

Mortimer and Co., coal merchants
John Mortimer, ale and stout mer.
Salt & Co's ales, J. M. Laney, agent
Mortimer & Clute, wine & spirit
   merchants
17 W. H. Jones, wine & spirit merchant

Foster's Chambers—

17 Edward M. Harwood, solicitor
17 Henry F. Lewes, junr. solicitor
17 Wm. Kent (tax office)
17 M. H. Bessell, tax office
17 A. F. Moreom & Tribe, stock and
   share brokers
17 W. H. Brown, solicitor
17 E. C. Barnard
   The Northam Assurance Company
   Assize Courts
Bristol School Board.—Clerk, Benj.
   Wilson
Bristol Law Library—Librarian,
   J. Thomas
   Wm. Tricks, Son, & Co, auctioneers
   —*Albion Chambers, which see*
   Thomas Guth, watch maker, etc
Exchange dining rooms company
   manager, Edward Carles

Small Street Court—

B. S. Stock, stock broker
Wm. Gregory & Son, solicitors
Cross and Co., trustees in bank-
   ruptcy, etc.
W. Hort, tax office
Wm. F. Brookman, accountant
Thos. Derrick, rate collector
William Derrick, rate collector
A. G. W. Jeffreds, accountant
United Loan & Discount Asso-
   ciation, Limited, D. M. Whit-
   tard, sec
5 C. S. Williams, hosier
26 H. Miles, watchmakers, etc

Sneyd Park. *See also Stoke Bishop.*

Matthew Kenyou - Stow, Sneyd
   park lodge
Robert Linton, Cairns villa
Alfred N. Price, Fern Hollow
E. G. Muller, Clarendon villa
John Fitch, Park villa
Joseph Watkins, 1 Elm cottages
J. L. Matthews, Blagdon house

here cross over

D. H. Walsh, Ivywell house
Miss Tuckett, Melbourne villa
Fred Henry Poole, Chester cottage
Rev. J. F. Wickenden, Stoke green
Christopher Hy., Ludlow, Buyclife
William Baker, Sneyd park villa
Jos. B. Taylor, Tudor lodge
Miss C. Strode, Algiers lodge
Col Jones Mortimer, Avonmoor
Wm.Moline,Locarno,1 SeaWall vils
Geo. W. Edwards,2 Sea Wall villas
J. H. Hirst, Avonhirst
Henry Oillaud, Avon grove
Dr. H. H. Goodeve, M.D. and J.P.
   Cook's folly
Dr E. Goodeve, Drinagh
Patrick Watson, Towerlenze
T. Y. Venn, the Oak
Peter D. Frankerd, The Knoll
James Derham
J. Colthurst Godwin, Clevelands
Elisha S. Robinson, J.P.
W. Pethick, Woodside
Edward P. Wills, Hazlewood
Mrs Col G. A. Harrison, Poulton lg
W. Walter Stoddart, Grafton lodge
Wm. Vachell, Severn house
Edward Heaton, Beaconsfield
Arthur Talmadge, Cecil lodge
Mrs Tharp, ladies' boarding
   school, Severn View
Mrs C. Tindal, Coral house
Dr J. F. Sheldston, Chelsfield
Christopher Godwin, Avonbac, J.P.
Robert Clarke, Towerhirt
J. R. Turner, Woodville
Arthur Henry Wansey, Senburgne

RESIDENTIAL MOBILITY

Road FERNBANK RD. District BRISTOL 6 Time Period 1880 to 1898

Year	House Number																				Change	
	2	4	6	8	10	12	14	16	18	20	22	24	26	28	30	32	34	36	38	40	Total	%
1880	O	O	O	O	O	O	O	X	X	X	O	–	O	O	O	O	–	–	O	O	3	
1881	O	X	X	X	O	O	O	O	O	O	O	–	O	O	O	O	–	–	O	O	3	
1882	O	O	O	O	O	O	O	O	O	O	O	O	O	O	O	X	–	–	X	O	2	
1883	O	O	O	O	O	O	O	O	O	O	O	O	O	O	O	O	–	–	O	O	0	
1884	O	O	O	O	O	O	O	O	O	O	O	O	X	O	O	O	–	–	X	O	2	
1885	O	O	O	O	O	O	O	O	O	O	O	O	O	O	X	O	–	–	O	X	2	
1886	O	O	O	O	O	O	O	O	O	X	X	O	X	O	O	O	–	–	O	O	3	
1887	O	O	X	O	O	O	O	X	O	O	O	O	O	O	O	X	–	–	O	O	3	
1888	X	O	O	O	O	O	O	O	O	X	O	O	O	O	O	O	–	–	O	O	2	
1889	O	O	O	O	O	O	O	O	O	O	O	O	O	X	O	O	–	–	O	O	1	
1890	O	O	O	O	O	O	O	O	O	O	O	O	O	O	O	O	–	–	X	O	1	
1891	O	O	O	O	O	O	O	O	O	O	O	O	O	O	O	X	O	O	O	O	1	
1892	O	O	O	O	O	O	O	O	O	O	O	O	O	X	O	O	O	O	O	O	1	
1893	O	O	O	O	X	O	O	O	O	O	O	O	O	O	O	O	O	O	O	O	1	
1894	O	O	O	O	O	O	X	O	O	O	O	O	O	O	O	O	O	X	O	O	2	
1895	O	O	O	O	O	O	O	O	O	O	O	O	O	O	X	O	X	O	O	O	2	
1896	O	O	O	O	X	O	O	O	O	X	O	O	O	O	O	O	O	O	O	O	2	
1897	O	O	O	O	O	O	O	O	O	O	X	O	O	O	O	O	O	O	X	O	2	
1898	O	O	O	O	X	O	O	O	O	O	O	O	O	O	O	O	O	O	O	O	1	

O — no change from previous year

X — change from previous year

— no entry

Figure 6.13 Sample form for use in a residential mobility study compiled from street directories

expanded and pressure built up on vacant sites in the older areas, much infilling took place, and stables were replaced by houses; when this happened all the house numbers could be changed within a year, and this makes the identification of some houses really quite difficult. One is reduced to using the names as a guide to how the renumbering was done. Then again when little villa developments were sold their new occupants often symbolised their new status by giving their house a name rather than a number. Often the local authorities later insisted on numbering and the transition from name to number can also cause some problems. It is not unknown for the name of a road to be changed and a contemporary street map is always a great help in interpreting an old directory.

When the details have been entered on the forms it is a simple matter to calculate, for example, the number of moves in any one year or from any one house over a period of years. Percentages need only be calculated for whole streets, unless there is good reason not to do so, for example, because there are several very distinct types of houses in a particular street. In this way one can see whether some houses, perhaps because of size, site or some other characteristic attracted a more or less permanent resident. One can also compare the frequency of moves from district to district, to see if certain districts have significantly greater turnover rates than others, whether there is a tendency for people in large detached houses to remain longer at the same address (having reached the higher rungs of the local social ladder) than those in the semi-detached villa districts where perhaps there was a greater degree of mobility, in whatever direction. Were the working class districts more or less stable in residential terms than the middle class districts? Were there any periods when there were much higher than average levels of mobility? Were some areas of small traders, shopkeepers and craftsmen more stable than others? Can one hope to complete some longitudinal profiles of individuals, tracing them from one address to another, perhaps even extending to a second generation if the names are sufficiently distinctive? Especial care is needed, if tracing individuals, so that one is not at some point confusing one individual with another and thus transposing the final part of their lives and seriously affecting the validity of the conclusions drawn.

As in so many other cases it is likely that any study along these lines will raise more questions than it answers, but for all the pitfalls and possible inaccuracies the directory method has the advantage of being reasonably quick and generating a large quantity of data. But it is best to start any investigation with a definite hypothesis, since there is no advantage to be gained from having data but no problem.

6.2.3 Possible topics from directories

Trade

Changes in the types of shops in a road or a small local shopping centre.

Ribbon development along a major radial road, with special reference to changes in the type and location of retail services.

Changes in the retail structure in or on the fringes of the central shopping area.

The corner shop, changes in distribution over time.

Changing patterns of consumer durable outlets, e.g. electrical, radio and TV shops.

Changing distribution of certain crafts and trades.

Locational changes in restaurants, public houses.

Development of facilities resulting from an innovation, e.g. garages and petrol stations for motor cars.

Distribution and growth, decline or general movement of a specific trade or profession in time.

Rate of change in ownership of certain businesses.

Residential and social

Changes in the patterns of different religious denominations.

Changes in the location and character of schools.

Residential mobility as revealed by changes in the names of occupants from year to year.

Contrasts in residential mobility between different areas.

Residential development and the growth of public transport in the late nineteenth century.

Location of recreational facilities, comparisons over time and/or space.

Changes in bus or tram services, both in terms of the changing network and changes in frequency of service.

The development of a new road, tracing changes in function and character from inception to the present.

Relationship between extension of residential development and the provision of schools.

Charting the annual spread of residential development in a suburb.

Changes in the number and location of police stations or post offices.

6.3 RATE BOOKS

6.3.1 Introduction

From a very early period it has been the custom to raise money for local administration from those who own property and usually the more valuable the property the higher the sum to be paid. In urban areas it was normal to assess each property by means of a valuation made every quarter. Details of each property, who owned it, who occupied it and how much rate was due were recorded in the rate books in the custody of the local authority (*Figure 6.14*). In 1925 the Rating and Valuation Act required all local authorities to preserve the books and to give the public full acess to them. Unhappily even by this time many of the older rate books had been lost, damaged or destroyed. Holmes (1973) in a detailed examination of the records of the town of Ramsgate was able to use no fewer than 305 volumes going back to 1803 without a single break, and with only a few gaps back to 1717. It is unlikely that many places will have still such a full record but even if there are large gaps, what has been preserved can be made to yield a great deal of valuable material.

In his study of Ramsgate, Holmes concentrates on two uses, first the study of property ownership, and second the extent of intercensal migration.

From 1837 the rate books list, for individual properties, the names of both owners and occupiers, but as usual there are difficulties in using the material and it takes patience and persistence to make the most of it.

6.3.2 Problems

First, it is necessary to identify the property. The order of listing in the rate book follows the route taken by the assessor and some houses he visited do not have names or numbers, some change numbers and names or from names to numbers, some even change from one street to another over time (this can easily happen with corner properties).

Second, the rating officers were not always as particular as they should have been about describing the property accurately. Commercial premises are often classed in a vague way as Office, Lock up shop, House and Shop. Comparison with contemporary street directories suggests that in many cases in the last century small private houses were being turned into shops by converting the front room and such premises were often rated as residences after the change had taken place.

Third, identifying ownership depends in some measure on having a high degree of differentiation of surnames; it would be difficult to trace owners, for example, if every other family were called O'Reilly, Mackintosh or Smith. It is also much easier if the settlement being studied is fairly small with a population of thousands rather than tens of thousands.

Table 6.2 shows that in Ramsgate in 1851 30 owners, representing under 5% of the total number of owners, controlled approximately 30% of the houses. This was the group that

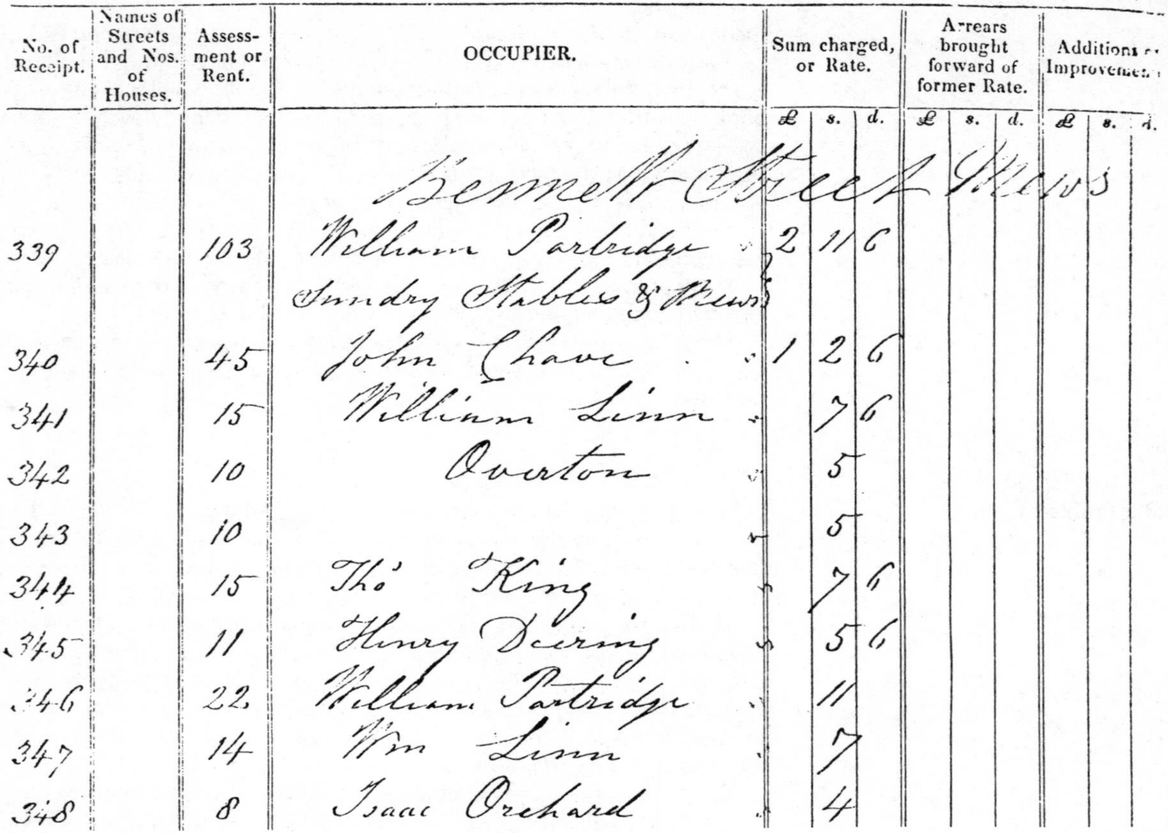

No. of Receipt.	Names of Streets and Nos. of Houses.	Assessment or Rent.	OCCUPIER.	Sum charged, or Rate.			Arrears brought forward of former Rate.			Additions or Improvements
				£	s.	d.	£	s.	d.	£ s. d.
			Bennett Street Mews							
339		103	William Partridge Sundry Stables & Mews	2	11	6				
340		45	John Chave	1	2	6				
341		15	William Linn		7	6				
342		10	Overton		5					
343		10			5					
344		15	Tho King		7	6				
345		11	Henry Deering		5	6				
346		22	William Partridge		11					
347		14	Wm Linn		7					
348		8	Isaac Orchard		4					

Figure 6.14 Extract from the Rate Book for the City of Bath, April, 1831 (By courtesy of the Bath City Archivist)

RECEIVED.			Folio in Cash Book.	Decision of Magistrates and of Churchwardens and Overseers, in respect to Persons applying for Relief, Time to Pay, &c.	Relieved, Appeals, and Poor.			Bad, Gone, Void, &c.			Arrears to be carried forward to next Book.		
Amount.													
£	s.	d.			£	s.	d.	£	s.	d.	£	s.	d.
2	11	6											
1	2	6											
	7	6											
						5							
						5							
	7	6											
	5	6											
	11												
	7												
	4												

131

owned 10 or more properties each. Most of them were local shopkeepers and builders. Half of the properties in the town were owned by small landlords with five houses or less each. Eighty per cent of all houses were held on lease; 26.3% of all houses were held by absentee landlords who did not live in Ramsgate, a high figure probably connected with the town's role as a holiday resort. The rate books are the only source of information of this kind which is crucial to our understanding of the way that the housing market worked in the period of rapid urban growth in the nineteenth century, and Holmes' study is of particular interest for this very reason.

TABLE 6.2 CONCENTRATION OF OWNERSHIP, RAMSGATE, MAY 1851. ALL FIGURES ARE CUMULATIVE (After Holmes, 1973)

	Properties	Owners	Properties (%)	Owners (%)
1 or more properties	1918	650	100.0	100.0
2 ''	1570	303	81.9	46.6
3 ''	1360	198	70.9	30.5
4 ''	1165	133	60.7	20.5
5 ''	1037	101	54.1	15.5
6 ''	932	80	48.6	12.3
7 ''	824	62	43.0	9.5
8 ''	747	51	38.9	7.8
9 ''	635	37	33.1	5.7
10 ''	573	30	29.8	4.6
15 ''	428	18	22.3	2.8
20 ''	289	15	15.1	2.3
25 ''	228	7	11.9	1.1
30 ''	173	5	9.0	0.8
35 ''	110	3	5.7	0.5
40 ''	40	1	2.1	0.1

6.3.3 Migration

Migration can also be studied by comparing the names of occupiers in one rate book with those in another. Making allowance for death, the extent of migration can be calculated. Holmes studied migration in Ramsgate between May 1851 and 1853, taking the same month in both years to offset seasonal variations in this resort town. Difficulties mainly consisted of the extra effort needed to search burial registers to find out how many of those missing from the later books had died. Names were a problem, especially with common surnames but since it was in the valuer's interest to avoid confusion when he was making his house-to-house inquiries it usually happens that middle names or initials were used to distinguish individuals or sometimes the suffix 'Junior' was added for the same reason.

In Ramsgate 10% of the occupiers migrated outwards in the two year period and 33.9% of the houses had a change of occupier in the same two year period. More detailed work suggests that turnover was often much higher, with tenancies sometimes lasting little longer than a quarter — 70% of the moves were by those occupying a house in Ramsgate for the first time. Chain moves were by no means uncommon and *Table 6.3* shows how details of such moves can be worked out by a careful study of the rate books. Holmes points out that the only casualty of this particular chain of linked moves was Thomas Hurst, who moved into lower rated property, whilst all the others, with the exception of Joseph Spratling who moved sideways, moved upwards into higher rated property. But in Ramsgate at this time downward moves were more common than this table suggests; in fact one quarter of all the moves were in a downward direction. The rate books can thus be used in studying social as well as geographical mobility.

TABLE 6.3 CHAIN OF OCCUPANCY CHANGES, RAMSGATE, MAY 1851 to MAY 1853, INCLUDING RATEABLE VALUES (RV) (After Holmes, 1973)

Name	Old address	New address	Old RV	New RV
Joseph Conder		21 Frederick Street		£10.50
Ebenezer Handford	21 Frederick Street	10 Plains of Waterloo	£10.50	£17.00
Thomas Hurst	10 Plains of Waterloo	9 Hibernia Place	£17.00	£11.50
Joseph Spratling	9 Hibernia Place	14 Hibernia Place	£11.50	£11.50
William Meader	14 Hibernia Place	5 Plains of Waterloo	£11.50	£13.00
Thomas Sturges	5 Plains of Waterloo	7 Camden Place	£13.00	£18.50
Sampson D. Campbell	7 Camden Place	4 Wellington Crescent	£18.50	£42.00

6.3.4 Conclusion

Holmes' conclusion was that the rate books are an extremely valuable source. The high turnover rates that he found in Ramsgate, which only further research will show to be typical or untypical, show that merely using census data, even at the level of the enumerators' returns, gives a gross underestimate of the amount of movement. Many more similar studies need to be undertaken before any general patterns begin to emerge and the laborious nature of the work suggests that for the individual researcher it is best to select a short time period and a small settlement. Only where the resources of a group are available should a larger settlement or a longer period be attempted, and then only after a pilot study has been conducted to test the data source and the methods thoroughly.

6.4 LOCAL NEWSPAPERS

The local newspaper came into its own as a vigorous forum for the reporting of local events and the discussion of local and national issues in the second half of the last century. Automatic printing machines, the effect of the electric telegraph and railways on communications, and the speed and ease of distribution, combined with the abolition in mid-century of a heavy stamp duty, allowed every important provincial centre to support at least one daily morning and several evening newspapers. Before long even small country market towns had their own weekly or twice weekly papers. The local paper became an important source of information with events being reported with an immediacy associated with eye-witness accounts. But newspapers are also useful for the incidental information they contain, that patience and imagination can glean.

Local newspapers can be used for tracing the stages of growth of a town. One sees reports of new building schemes, advertisements for new houses, notices about changes in tram and, later, bus routes following urban extension. Disagreements about difficult or controversial developments appear in the paper and readers' letters often give some flavour of public attitudes to affairs of civic importance. It is possible through property advertisements to get some idea about the changing status of different areas in the town, and about changing property values. The progress of Public Health legislation is revealed by reports of local investigations, of the progress for example of new sewerage schemes, road improvements and drainage schemes.

Industrial changes are reflected in the reports of openings, extensions and closures of local factories. From the beginning advertisements have been an important source of revenue for the local newspapers and so of information for the enquiring student.

Most newspapers keep a complete set of their back numbers. The earlier volumes are usually bound and where there may have been gaps these are sometimes filled by microfilmed copies of the missing issues. Local reference libraries also keep back numbers and for ease of use the reference library is generally more comfortable than the local newspaper office.

It is easy to be deflected from one's main purpose when one starts browsing through back numbers of old newspapers and most people find that they underestimate the time it will take to accomplish a particular task.

One very interesting use of local newspapers is in determining the area over which they circulated. This is easier the more recent the period being studied. The technique is simple but tedious. It involves sampling a small number (at random) from say a year's run of a particular paper. All that is necessary is to plot on an appropriate scale map all the places in the vicinity to which reference is made, in news items or in advertisements. The end result is a map with a large scatter of dots, the most remote of which, linked up, define the outermost limits of the area over which the paper probably circulated in significant numbers. The assumption is that news would only be reported from places where readers of the paper lived. The method is a little crude, the more so the more remote the period concerned. Where there are plenty of social and sporting events to report, and a reasonable number of small advertisements the accuracy improves. Even so it is likely that there will be quite an area of overlap between neighbouring newspaper circulation areas. In many country areas the same basic newspaper circulates over a wide area but for each market town there is a distinctive title and a page of news relating to that particular vicinity, though much of the editorial and feature material is the same for all editions. In that sense the different newspapers are not strictly speaking competitive. On the other hand even in small towns there was often more than one weekly newspaper, perhaps coming out on different days of the week, and to concern oneself with only one of them would perhaps be to give an unbalanced and inaccurate picture.

REFERENCES

Directories

Davies, W.K.D., Giggs, J.A., and Herbert, D.T. (1968). 'Directories, Rate Books and the commercial structure of towns.' *Geography* **53** (Pt 1), 41-54

Goss, C.W.F. (1932). *The London Directories 1677—1855.* London; Denis Archer

Norton, J.E. (1950). *Guide to the National and Provincial Directories of England and Wales Published before 1865.* London; Royal Historical Society

Oliver, J.L. (1964). 'Directories and their use in geographical enquiry.' *Geography* **49** (Pt 4), 400—409

Wild, M.T. and Shaw, G. (1974). 'Locational behaviour of urban retailing during the nineteenth century, the example of Kingston upon Hull.' *Trans. Inst. Br. Geogr.,* No. 61, March, pp101—117

Rate Books

Holmes, R.S. (1972). 'Identifying nineteenth century properties.' *Area* **6**, 273—77

Holmes, R.S. (1973). 'Ownership and migration from a study of rate books.' *Area* **5**, 242—251

CHAPTER 7 TRANSPORT

7.1 ROADS

Material relating to the history of roads is found in a wide variety of sources, too numerous to list in full.

Some have already been mentioned in earlier sections. For example, tithe redemption maps and maps accompanying enclosure awards give good local detail, and the 'General Views...' of the Board of Agriculture reports are useful at the county scale. Parish records, particularly the accounts of the highways' surveyors and the vestry minutes, cover the period from 1555 to 1835 when the parish was responsible for the upkeep of its own roads. Between 1836 and 1888 the responsibility for the roads was largely shifted on to larger highways boards whose records are usually to be found in the Quarter Sessions proceedings. After 1888 such records were kept by the County Council. Directories, as we have seen, give details of coach and carrier services and later of motor bus services. Turnpiking of the more important roads took place from the late seventeenth century, and since all such improvement required an Act of Parliament there is often some local copy of petitions that went to the appropriate parliamentary committee to support the Bill. Local newspapers often printed the prospectus issued by the aspiring proprietors of the Turnpike (and also canal)Trust set up to administer the new development. The most complete geographical account of the turnpiking is by E. Pawson (1977). This traces the diffusion of the turnpike system through time and space and assesses the economic impact of this technical innovation. It is invaluable as a background against which the local experience can be viewed.

The records of tramway and bus companies, where they have been preserved, are also useful, but so many of the early bus companies in particular were so small and their records so fragmentary that even if they have survived successive amalgamations they are often tantalisingly incomplete. If you are studying the changing accessibility of settlements over time it is usual to find difficulties in the period after 1914. Earlier, the carriers are fully represented in the directories. But after 1914 small private bus operators flourished in a strongly competitive situation. Many later went out of business and were absorbed by more successful enterprises. The formation of the national companies in the inter-war period spelled doom for many of the small operators, but in some remote areas a number continued to survive, offering market day scheduled services and making up their income by private hire for social and sporting occasions. It is these small operators who are sometimes hard to pick up from the printed record. More recently all stage carriage operators have to be licenced by the Regional Traffic Commissioners and so they can be traced.

The geographer is most likely to be interested in the roads as links in a communication system. Historically the opening of new and the closing of old links, by altering the characteristics of the network, may reflect or cause changes in the character of settlements. There is much scope for medium-scale studies of the relationship between changes in the nature of the road network and changes in the function of settlements connected by that network. A scale larger than a parish is appropriate for such a project. Measurement of network characteristics is straightforward. More problems arise in trying to categorise settlements on the basis of function. In preliminary studies it is often satisfactory to choose one characteristic of the settlement as a surrogate for many others. The easiest characteristic to choose is population. Generally the larger the settlement or the faster it is growing the more vigorous its economy is likely to be. Places that become linked with important nodes in an expanding road network are likely to grow more rapidly than places that remain relatively isolated, or so at least we may hypothesise.

7.2 CANALS

Canals threaded a surprisingly large part of the British countryside in the early part of the last century and there are many parishes through which the often abandoned sections of once busy waterways now lie almost unrecognised. It would be a great mistake however to suppose that these canals have gone unrecorded. The great majority of them have been the subject of thorough investigation by historians and canal enthusiasts and there is an extensive literature on many aspects of their construction, operation and economics. The recent revival of interest in restoring sections of canals has attracted considerable public sympathy and some measure of official support, and in many areas there are active preservation and restoration societies. As a general rule it is wise to begin with the assumption that most of the information about any part of the canal system is already available, and to seek it out by a diligent search in the local library. The references on pages 143—144 may help initially; then it is a matter of following up other references within these in order to get more detail.

From a strictly geographical point of view, details of the history of a canal are only of incidental interest, in that they may possibly illuminate some aspects of geographical importance. If the central point of concern is, let us say, the changing geography of a parish through which a canal happens to pass, the canal may be only of the slightest interest, perhaps because it brought quantities of building material and above all of lime to improve the quality of the local soils. Canals did this for nearly all the places through which they passed. But the canal might be of critical importance for the parish if, through its presence, it created employment, perhaps through trans-shipping facilities (Porteous, 1977) or the need at that point for a pumping station or for a flight of locks to be operated and maintained. In such a case it would be proper, indeed essential, to study the canal in some detail. But if its impact were small then it need attract no more comment than the passage today of high voltage overhead cables. Unless they become for some reason a local issue they deserve no more than a passing mention.

If on the other hand the geographer's concern is in the canal as part of a wider regional transport system, then the reasons for the construction and decline of the canal, and the characteristics of the network of which it was a part, are of central concern and the historical perspective is desirable.

One should always make sure that there is a valid geographical reason for taking a detailed interest in a particular canal.

In certain parts of a canal the lie of the land, the geology, the need to keep a supply of water at summits to make good losses through locks, the search for the cheapest and best way to route the canal over watersheds between river basins, all may be most important local considerations, with a strong physical emphasis, and very much the proper concern of the geographer.

Farrington (1973) in an unusual study shows a distinctively geographical approach to canals. He studied 11 of the main trunk canals of England and considered the relationships which existed between canal routes and canal morphology. He emphasised the physical influences through field work and used morphological indices to analyse the relationships between canal and land form.

The morphological indices he developed were (a) an index of sinuosity, and (b) a mean figure derived from the height of each embankment and the depth of each cutting. The latter index is probably self-explanatory but the sinuosity index may require a fuller description.

The sinuosity index measures the degree to which the canals deviate from a straight line on a small scale. Obviously for a canal joining two places the shortest route, a straight line between the two places, would usually be totally impracticable and if we were to measure the sinuosity of the entire course between the two places the result would not be very meaningful; some of the deviation may have been to collect more traffic from intermediate points and not because of topographic constraints so Farrington devised the following method.

First it is necessary to work with a large-scale map and the O.S. 1:25 000 maps are ideal. Suppose the canal in a certain stretch has the form shown in **Figure 7.1**. First of all, points along the line of the canal are marked off at constant intervals from each other,

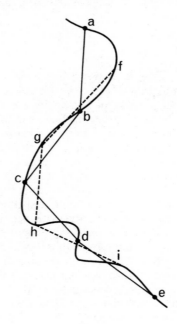

Figure 7.1 Measurement of sinuosity

say in this case at one kilometre intervals as if they marked the position of distance markers along the bank of the canal itself. In the diagram these points are marked *a,b,c,d* and *e*. Then intermediate points are added midway along each unit length, again marked along the course of the canal. These intermediate points are marked as *f,g,h* and *i*. So along the course of the canal the succession and spacing of the points is *af=fb=bg=gc=ch=hd=di=ie*.

The sinuosity value of canal length **bc** is then given by:

$$\frac{1}{2}\left(\text{straight line } \textbf{\textit{bc}} + \frac{(\text{straight line } \textbf{\textit{fg}})}{2} + \frac{(\text{straight line } \textbf{\textit{gh}})}{2} \right)$$

Similarly the sinuosity value of canal length **cd** is given by:

$$\frac{1}{2}\left(\text{straight line } \textbf{\textit{cd}} + \frac{(\text{straight line } \textbf{\textit{gh}})}{2} + \frac{(\text{straight line } \textbf{\textit{hi}})}{2} \right)$$

In this way the sinuosity value for each stretch is computed, each value representing the sinuosity of one kilometre of canal. If the canal is absolutely straight the sinuosity value will be the same as the interval between the two points, i.e. if the scale is 1:25 000 and the interval every kilometre (4 cm on the map) then a sinuosity value of 4.00 would indicate absolute straightness. In the example given from Farrington he used one mile as his interval and with the 1:25 000 map this is the same as a scale of 2½ inches to one mile. In this case therefore the interval between the points was chosen as one mile (2½ inches) and so an absolutely straight stretch would have had a sinuosity value of 2.5. The lower the sinuosity value below the value of a straight stretch, the greater the degree of sinuosity.

The sinuosity of an entire canal can be expressed either as a mean figure, a mean that is of all the values for each kilometre stretch, or as a histogram with each bar representing a particular value. Such a form of analysis is well suited to an investigation of an entire canal or to a comparison between different canals but it would be of little relevance over a short stretch of a single canal. *Figure 7.2* shows an example of the method used, and it is clear how the three characteristics plotted are interrelated. The higher the sinuosity values the straighter the course. The very low values between the 28 and 34 mile markers represent the need for the canal to maintain its height for some distance in order to secure a water supply at Gailey. To maintain the level the canal had to follow a sinuous course along the contour. On the whole, rapid changes of level through lock construction lead to relatively straight sections and therefore to high sinuosity values.

The most obvious feature of the cutting profile is the deep cutting that occurs between Stourport and Swindon, and apart from the actual summit the complete absence of any cutting more than 15 feet deep on the rest of the canal. This stretch of deep cutting was caused by the narrowness of the Stour valley. In order to avoid the river and the string of water-powered mills along its course the canal had to be cut into the steep valley sides.

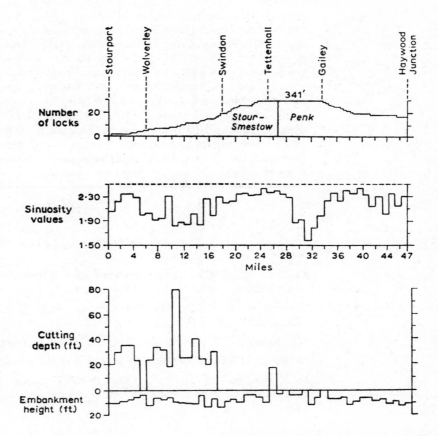

Figure 7.2 Morphological profiles of the Staffordshire and Worcestershire Canal (After Farrington, 1973)

7.3 RAILWAYS

The railways succeeded the canals as a major influence on industrial, commercial and urban development. As with the canals scarcely any stretch of railway is without its chronicler. Even the minor but locally important tramways and early railroads of the mining districts have been the subject of detailed investigation. Often the flavour of such studies is more technical or historical than geographical, but even so local libraries should be searched thoroughly before deciding on a work programme. There are so many books and booklets on railways, from the definitive histories of the main companies (for example, E.T. MacDermott's three volumes on The Great Western, revised by C.R. Clinker in 1964) to short but informative pamphlets about narrow gauge mineral lines, that there is little to be gained by giving a selective list of them. There are two major bibliographic references that should be consulted if possible if a serious effort is to be directed at combing the relevant literature about a particular railway. *A Bibliography of British Railway History* (Ottley, 1965) is elaborately classified and comprehensive. *British Transport: an Economic Survey from the Seventeenth Century to the Twentieth* (Dyos and Aldcroft, 1969), as its title suggests, covers roads and canals as well as railways.

REFERENCES

Pawson, E. (1977). *Transport and Economy. The Turnpike Roads of Eighteenth Century Britain.* London; Academic Press

Porteous, J.D. (1977). *Canal Ports, the Urban Achievement of the Canal Age.* London; Academic Press

Farrington, J.H. (1973). 'Morphological studies of English canals.' University of Hull, Occasional Papers in Geography, No. 20

MacDermott, E.T. (revised by C.R. Clinker) (1964). *The Great Western.* London; Ian Allen

Ottley, G. (1965). *A Bibliography of British Railway History.* London; Allen and Unwin

Dyos, H.J. and Aldcroft, D.H. (1969). *British Transport: an Economic Survey from the Seventeenth Century to the Twentieth.* Leicester University Press

FURTHER READING

Canals of the British Isles series. Newton Abbott, Devon; David and Charles

Boyes, J. and Russell, R. *The Canals of Eastern England*

Delany, V.T.H. and Delany, D.R. (1966). *Canals of the South of Ireland*

Hadfield, C. (1970). *British Canals: An Illustrated History,* 5th Ed

    ''   ''   (1970). *Canals of the East Midlands, Including Part of London*

    ''   ''   (1969). *Canals of South and South East England*

    ''   ''   (1967). *Canals of South Wales and the Border*

    ''   ''   (1967). *Canals of South West England*

    ''   ''   (1967). *Canals of Yorkshire and North East England*

Hadfield, C. and Biddle, G. (1970). *Canals of North West England*, 2 Vols.

Lindsay, J. (1968). *Canals of Scotland*

McCutcheon, W.A. (1965). *Canals of the North of Ireland*

Inland Waterways Histories, Edited by C. Hadfield. Newton Abbott, Devon; David and Charles

The Ballinamore and Ballyconnell Canal by P. Flanagan

The Birmingham Canal Navigation by S.R. Broadbridge

The Bude Canal by H. Harris and M. Ellis

The Dorset and Somerset Canal by K.R. Clew

The Grand Canal of Ireland by R. Delany

The Grand Junction Canal by A. Faulkner

The Grand Western Canal by H. Harris

The Great Ouse by D. Summers

The Kennet and Avon Canal by K.R. Clew

The Leicester Line by P.A. Stevens

London's Lost Route to Basingstoke by P.A.L. Vine

London's Lost Route to the Sea by P.A.L. Vine

The Nutbrook Canal by P. Stevenson

The Royal Military Canal by P.A.L. Vine

The Somerset Coal Canal and Railways by K.R. Clew

The Thames and Severn Canal by H. Household

The Yorkshire Ouse by B.F. Duckham

The Oxford Canal by H. Compton

GENERAL REFERENCES

Apart from the sources mentioned at the end of each chapter there are several that are particularly useful for the breadth of the topics that they span and the richness of the sources that they mention.

Hoskins, W.G. (1972). *Local History in England,* 2nd Edn. Harlow, Essex; Longman

Stevens, W.B. (1973). *Sources for English Local History.* Manchester University Press

West, J. (1962). *Village Records.* London; Macmillan

Two periodicals *The Local Historian* and *The Amateur Historian* are also well worth consulting.

GLOSSARY

ADVOWSON	The right of patronage, or of presenting a clergyman to his living.
AFTERMATH	Herbage remaining after harvest.
AID	A periodic payment made by a feudal vassal to his lord.
ALIENATION	The transfer of land from one holder to another.
APPROPRIATOR	An ecclesiastical body or person, other than the parish priest, having the right to (usually) great tithes in the parish.
ASSART	Land cleared of wood and rendered arable. To assart land in the forest without a licence was a grave offence for whereas waste of the forest consisted in cutting down trees that might grow again, assart was rooting them up so that they were forever destroyed. Where ground was assarted by licence from the Crown, such ground was subject to assart rents.
BALK	An untilled boundary strip, usually between two selions.
BEREWICK	Subsidiary or outlying estate.
BOVATE	The law Latin equivalent of the English term 'oxgang'.
BRECKS	Supposedly temporary enclosures which in fact often became permanent ones, especially from forest lands, c.f. intakes.
CALENDAR	A list of documents with summaries of their contents.
CARR	Common, especially marshy common.
CARTULARY	A list or record of lands and privileges granted by charter.
CATTLEGATE	The right to graze a single beast on land, in the soil of which one has no legal interest.
CLOSE	A hedged or fenced or walled piece of land, often wrongly called a field.
COMMON LANDS	A term usually applied now to manorial waste land but applicable also to open fields over which common pasture rights existed.
COPYHOLDER	A tenant holding his land by copy of court roll, and belonging to a class deriving from the medieval villeins.

CORN RENT A cash annuity, varying in amount according to the price of corn in the area, hence a safeguard against any future change in the value of money.

COTTAGER The occupier, sometimes also the owner, of a tenement often having attached to it a croft, and usually a common right and a little land.

COURT BARON The lord's court held in and for his manor, and having jurisdiction in many matters of local concern. The court to which the free man of the manor owed suit.

COURT LEET A local court authorised by Royal grant to hear cases of petty jurisdiction. Such courts were often responsible for the view of frankpledge.

COURT ROLL The record of a court's activities, so called because the parchment on which the record was written was filed as a roll.

DEMESNE That part of the manor lands which the lord had not granted out, but retained in his hands. The early equivalent of the home farm today.

ENCLOSURE The conversion by any means, legal, extralegal or illegal, of common lands, arable, meadow, pasture or waste, into individual ownership, tenancy and use.

ENGROSSING The accumulation in the hands of one man and his family of agricultural holdings adequate to the maintenance of more than one family.

ESCHEATOR The royal officer responsible for holding Inquisiciones Post Mortem.

ESSART *See* assart.

ESTOVERS Properly signifying nourishment or maintenance. Common of estovers was the right of the tenant in a manor to take wood necessary for the use of his farm or house from his lord's estate.

EXTENT A detailed valuation of land and property.

FALLOW Land ploughed and harrowed, but left uncultivated for a year so that it may recover its fertility.

FARM Essentially a landholding let at a rent, or the rent or service paid for a property.

FEE SIMPLE Land held in fee simple was held without restriction on inheritance.

FIELD A large area of open land, arable, divided into furlongs each again subdivided into selions, all the selions in a furlong normally being subject to the same crop rotation.

FINE	A fee or an end. Often applied to a final agreement or decision concerning landholding. Such fines were written in triplicate, copies went to the disputants, and the third copies, feet of fines, were filed in the records of the court of common pleas.
FINE ROLLS	Records of payments made for writs issued by the crown.
FRANKPLEDGE	The system by which the householders of a manor or a village were grouped into tithings in order that each tithing could be held corporately responsible for the good behaviour of its members. Cases of lawbreaking were heard twice a year at a view of frankpledge.
FURLONG	1. A measure of length equal to 220 yards which was a convenient length for plough oxen to travel on an average soil, before resting and turning on the headland. 2. A subdivision of an open field. A block of continuous selions, all more or less the same length and all running the same way.
GELD	Payment or tribute. The tax paid to the Crown by English landholders before the Conquest and continued under the Norman kings.
GELD ACRE	(or hide) An acre or hide as reckoned for the purposes of geld.
GORE	A more or less triangular scrap of land usually in between two neighbouring furlongs.
HADE	Balk.
HALF-YEAR LAND	Lammas land, common land, arable or pasture occupied in severalty for part of the year, but after the crop was taken, pasturable by the occupants and others by agreement. Usually opened for grazing on Lammas Day (1 August) or Old Lammas (12 August).
INDENTURE	An agreement written two, three or more times on a single parchment. To insure against forgery the copies, one for each party to the agreement, were separated by irregular wavy cuts ('indents').
INQUISITION	An inquiry. Inquisicion post mortem was an inquiry into the holdings, services and succession of a deceased person, who held land of the king.
KNIGHT'S FEE	Originally a grant of land in exchange for undertaking to supply a lord with the services of a fully armed knight and his necessary servants for 40 days each year. Eventually payment was made by rent, and knights' fees were often divided among several tenants.
LAWN, LOON	(in Dorset mainly) selion.
LETTERS CLOSE	Private letters.

LETTERS PATENT	Open letters.
LEY	Land ploughed for a year, then left to grass for another period sometimes of several years.
LOT MEADOW	Common meadow land in which the mowing rights are distributed each year by an arrangement having some degree of chance in it, often a primitive form of lottery.
MANOR	A feudal freehold estate. Manors were always held in fee simple so they passed automatically from the lord to his heir.
MARK	13s. 4d.
MARLING	Adding clay to a light soil to improve the texture.
MARSH	Often a common, not always a quagmire, sometimes suitable for grazing without extensive drainage.
MASLIN	Mixed corn, properly it should be a mixture of wheat and rye. Sometimes grown as a mixed crop, sometimes mixed after thrashing.
MEAR, MEER	A boundary balk.
MESSUAGE	Not merely a dwelling house, but essentially a farmhouse with land annexed to it.
METES	Boundaries.
MOIETY	One of two parts into which an estate was divided, not necessarily a half.
NOVEL DISSEISIN	Judicial cases concerned with dispossession of land.
OYER AND TERMINER	To hear and to give judgement. A court of oyer and terminer was one of final judgement usually held by one of the king's judges.
OXGANG	Half as much as a yardland, since two oxen went to a yoke one-eighth of a ploughland. On a very rough average about 15 acres.
PANNAGE	The right to feed pigs in the woods. Also the payment made to hold that right.
PINDER	Manorial or parish officer in charge of the pinfold (pound) and the detention therein of straying stock.
PLOUGHLAND	As much open arable land as would occupy throughout the year an eight-ox plough team, hence a conventional unit of area, varying widely from county to county and parish to parish with soil conditions but always containing four yardlands, i.e. eight oxgangs. A reasonable average figure is 120 acres, putting it in the same category as the Domesday Hide.

PURPRESTURE	An encroachment, especially on to deer pastures in the forest.
QUIT RENT	Money payment made by a smallholder in place of traditional services.
SCUTAGE	A money payment made to the king instead of personal military service. Scutage became increasingly usual from the reign of King Henry II onwards.
SEIZE	Possess. A feeholder was said to be seized of his land. To be disseized was to be dispossessed.
SERF	An unfree servant; a slave.
SHOT	Furlong.
SOCAGE	Tenure without servile obligation; the tenant usually paid a rent.
SOKEMAN OR SOCMAN	A free tenant who came under the lord's jurisdiction.
SQUIRE	Technically a man of blood and coat armour, two degrees above the commonalty (the grades being gentleman, esquire, knight). In fact, the principal landowner in any rural village, provided that he or his family had been there long enough to establish a customary right to the title. Often the ultimate successor to the medieval lord of the manor.
SURCHARGING (the common)	Overstocking the common by turning on it more beasts than it could feed or than one was entitled to (sometimes referred to as overpressing).
TENANT-IN-CHIEF	One who held land directly from the crown.
THEGN	An Anglo-Saxon retainer of noble birth. By 1086 however many thegns owned little land and enjoyed few privileges.
TITHING	Originally a tithing had 10 members but eventually numbers varied considerably.
TOFT, TOFTSTEAD	Originally a homestead, the site of a house and its outbuildings, with its attached land and common rights. Hence a farmhouse having full common rights.
TOWNSHIP	The Old English unit of settlement, often the basis of the later ecclesiastical unit, the parish, or the manor.
VESTRY (meeting)	The ancient customary governing body of a parish which all ratepayers might attend, in order to deal with most matters of local concern, whether civil or ecclesiastical. Though it still exists it has lost nearly all its powers to the district or county authorities.
VICAR	Parish priest receiving only small tithe (now tithe rent charge) from his parishioners.

VILL	The Norman French equivalent of the Old English township.
VILLEIN	The increasingly unfree but landholding countryman of early feudal times.
VIRGATE	The Low Latin equivalent of the English yardland, hence one-quarter of a ploughland or approximately 30 acres.
VISITATION	Official inspection of the parish by or on behalf of the bishop or his archdeacon.
WAYWARDEN	Surveyor of the highways, parochial official having responsibility for the maintenance of the highways in the parish and the supervision of the labours of his fellow parishioners in their statutory obligation to maintain and repair them.
WONG	Furlong in the open fields or the name given to an enclosed meadow.
YARDLAND	(Virgate) As much land as would keep busy throughout the year a pair of oxen (in a two-ox plough).

INDEX

2374

Morgan, Michael
Alan.

Historical sources
in geography

DATE			